Electrical Installations in Building

VOLUME II

ER. HARI MOHAN JOHRI

KNOWLEDGE WORLD

KW Publisher Pvt Ltd
New Delhi

KW Publishers Pvt Ltd
4676/21, First Floor, Ansari Road, Daryaganj, New Delhi 110002
E knowledgeworld@vsnl.net T +91.11.23263498 / 43528107

w w w . k w p u b . i n

Second Edition 2010

ISBN 978-93-80502-47-2

Published by Kalpana Shukla, KW Publishers Pvt Ltd
4676/21, First Floor, Ansari Road, Daryaganj, New Delhi 110002

Printed and bound in India by Bhavish Graphics.

TEXT BOOK OF
UTTRAKHAND POLYTECHNIC

Contents

FOREWORD   v

PREFACE   vii

CHAPTER I: **Air Conditioning**   1

CHAPTER II: **Lightning and Its Protection**   41

CHAPTER III: **Fire Detection and Protection System**   71

CHAPTER IV: **Sound System**   104

CHAPTER V: **Stage Lighting**   129

CHAPTER VI: **Diesel Generating Set**   141

CHAPTER VII: **Lift**   157

CHAPTER VIII: **Computer Networking**   177

Prof. Mohibullah

Ph.D., M.Sc. Engg., B.Sc. Engg.
MIEEE, FIETE, FIE, MIEE,
MSSI, MISTE, MICTP

CHAIRMAN

Ref. No. ....

Dated: ....28.7.2010........

DEPARTMENT OF ELECTRICAL ENGINEERING
Z.H. COLLEGE OF ENGINEERING AND TECHNOLOGY
ALIGARH MUSLIM UNIVERSITY, ALIGARH - 202002

Phone: External	-	0571-2721178
EPABX	-	2700920-26
Extension.	-	3181 (Ch), 3180 (O)
Fax:	-	+91-571-2721178
E-mail:		chairman_eed@rediffmail.com

FOREWORD

I highly admire the author, an alumini of this University and graduate engineer of this department, for his contribution of a good text on "Electric Installation in Buildings". He is personally well known to me. It is a reflection of his long field experience in design, erection and estimation of various electrical installations for modern buildings.

Writing a text is indeed a very challenging task. Besides possessing the field experience in the subject, the presentation is in such a way that students can understand easily.

The book is very useful for the engineers and architects, specially for preparation of detailed projects reports (DPR).The first volume of this book is already in the market, covering internal and external illumination, air circulation, wiring network of appliances used in buildings, over head lines, electrical earthings etc.

I am sure that teachers and students at Diploma level and architecture students at college level will find this text much useful on the subject. The effort of the author is well appreciated and I wish him a great success in his life.

Mohibullah

Prof. Mohibullah

Chairman

Preface

The electric installation play vital role in the utilization of building, constructed for different use, e.g. residences, offices, hotels, shopping complexes, theatres, sport stadiums, auditoriums, especially multi-storied buildings.

The basic electrical installations are, lighting i.e. providing illumination both inside and outside buildings exhaust fans, use of portable and non-portable electrical machines or appliances and their wiring network, including sub-main wiring, cable, O.H. lines etc, including control panel and switches. The earthing is vaery common and essential electrical installation. All these basic installations have been covered in Volume I of this book.

The other electrical installations like air conditioning, various sound systems, protection against lightning and fire, lift, diesel generating sets, computer networking are various optional installation in various buildings. Protection against lightning and fire are mandatory in buildings as per building manual. Stage lighting, sound systems are essential in building used for various conference hall, auditorium, places of worship, studios and audio video broadcasting stations. Telecommunication and networking has become very useful electrical installation now-a-days.

The book describes these optional electrical installations necessary for the buildings and useful for occupants. Lift is useful for accessing high floors and shifting of essential commodities. D.G. sets are essential for alternate source of energy at time of failure of the power supply from the powers stations.

All these electrical installations have been discussed completely in this book Volume II, to help architects, engineers associated with building projects, their construction maintenance and to design estimating and costing.

Volume I was already published in 2008, covering all possible basic electrical installations in buildings and other optional electrical installations in buildings are being mentioned in this book "Electrical Installations in Buildings" Volume II. These installations with brief introduction, their classification, installation, operation, estimation and costing have been described in separate chapters of this book.

The author has written this book in the from of a text book for students, studying electrical engineering at polytechnics and also students of architecture to provide in-depth understanding on estimating and costing.

The author is thankful to, the manufacturer of the plants and components of the different installations, their authorized dealers and various agencies, involved with construction of these installations, who supplied literature and relevant broachers to author to help him for writing this book.

The author dedicates this book to his parents and in-laws as for inspiration. The book is also dedicated to engineers associated with buildings and engineering students who will find this book useful. Last but not the least the book is dedicated to my dear wife, without whose untiring love and support this would not have been possible at all.

The author will appreciate any suggestions and feedback if referred to him.

Hair Mohan Johri
FIE Mumber of IRE
Chief Engineering E/M
Uttar Pradesh Public Work Department

Air Conditioning

1.0 Introduction

Excessive air temperature during summer and excessive low temperature in winter affect adversely on the working efficiency of the human being. Increased temperature of atmosphere increases the temperature of body of a human being in summer. It not only disturb the function of different organ of body, but creates shortage of water inside body which goes out of body as sweat. This is why man feels thirsty and drink sufficient water. With increase of temperature air gets heated and rises above, to increase air density near ceiling of building or room and creates shortage of air or oxygen necessary for breathing near floor. Thus man feels quite uncomfortable in summer with excessive temperature rise. Increased humidity in air also gives uncomfort during rainy season. Similarly during winter, the atmospheric temperature decreases which in turn reduces body inside temperature and body heat goes out, causing cold and shivering and thus causes harm to human. For such cases of excessive high or low temperature, the air conditioning is quite useful to control the temperature & humidity to the desired value to give comfort to the human being during summer or winter.

1.1 Important Definition:

1.1.1 Heat: It is defined as the energy transferred, because of a temperature difference between a system and the surrounding, without transfer of mass across the boundary of a system. Its unit is J or KJ.

1.1.2 Sensible Heat: When a system is heated and the temperature rises as heat is added, the increase in energy is called sensible heat, $Q = mc(T2 - T1)$. Similarly heat removed and extracted, when it is cooled and temperature falls, is also called sensible heat.

1.1.3 Latent Heat: All pure substance can change their state i.e. solid, liquid or gas. The change of these state occur at same temperature and pressure. The heat required (added or removed) to bring the change with no change in temperature is called latent heat. Heat is added in case of Latent Heat of vaporization, while heat rejected in case of Latent Heat of fusion or condensation.

1.1.4 Dew Point: The temperature at which the water vapor in air is maximum and starts condensing, is called dew point. Water vapor density in air is maximum, when air is saturated, the air temperature at this stage is known as dew point.

1.1.5 Humid & Dry Air: If the water vapor is present in air, the air is termed as humid air and if there in no water vapor in air, the air is called dry air.

1.1.6 Humidity: This is a measure of water content of water vapor present in humid air. Its unit is grain / m³.

1.1.7 Absolute Humidity: Water vapor content per unit volume of air is called absolute humidity.

1.1.8 Relative Humidity: Ratio of real density of water vapor in air to the density of water vapor in saturated air.

1.1.9 Dry Bulb Temperature (D.B): The air temperature measured by a general D.B. thermometer is called dry bulb temperature. Thermometer is covered to protect it from moisture or radiation.

1.1.10 Wet Bulb Temperature (W.B): The air temperature measured with D.B. thermometer with bulb kept in wink (wet cloth bag) is called W.B. temperature is always less than D.B. temperature.

1.1.11 Fundamental Units: Every quantity is measured in terms of some arbitrary but internationally accepted units i.e. unit of mass, time, temperature etc.

1.1.12 Derived Units: The units of some quantities are expressed in terms of derived unit which are derived from fundamental units e.g. area, velocity, acceleration, pressure etc.

1.1.13 Isotropic Process: The process carried out at constant pressure is called isotropic process.

1.1.14 Isenttropic Process: The process performed with change in temperature is called isentropic process.

1.1.15 Throttling Process: The process in which some restriction is created in flow of a substance is called throttling process.

1.1.16 Enthalpy: The enthalpy of the substance is heat energy per unit mess of the substance.

1.1.17 Entropy: The change in enthalpy with change of one degree temperature of the substance.

1.2 Refrigeration

Refrigeration is a process of continued extraction or removal of heat from a place or any substance when temperature is already below the temperature of surrounding. So it is process of cooling or lowering temperature of any body/ substance below the temperature of its surrounding. The refrigeration primarily depends on two things (a) Refrigerants (b) Process.

(A) **Refrigerant:** A refrigerant is heat carrying medium which during their cycle in the refrigeration system absorbs heat from a low temperature system and discards heat to a higher temperature system.

(B) **Process:** The refrigeration is carried out by following processes of refrigerant (a) compression (b) expansion (c) evaporation (d) absorption.

1.2.1 Removel of Heat: There are two methods of heat removal.

(i) By Evaporation or Melting: When a liquid evaporates or solid melts it absorbs latent heat from surroundings, resulting cooling of surroundings e.g. when a man perspire during summer, the perspiration water evaporates on the body of man into atmosphere by absorbing latent from the body of the man through skin, so man feels cooling effect.

(ii) By Heat Transfer: The other method of removing heat is by heat transfer. During the process of traveling of heat energy from hot substances or space to cold substances or space, the hot space cools down and cold space heats up. This process is called heat transfer. It may be carried out in two forms (i) Directly (ii) Indirectly.

(a) Direct Heat Transfer: When the substance to be cooled is directly in touch with the colder substance e.g. ice cube when added in a glass of water.

(b) Indirect Heat Transfer: It is accomplished, when heat transfer takes place between two substance through a third substance e.g. Exhaust fan in room cooler gives cooled air to the occupant of the room in which cooler is located, so occupant reject heat through third substance air.

Further heat is transferred by a substance in three ways namely (a) Conduction (b) Convection (c) Radiation.

1.3 Refrigerant

The natural ice is a mixture of ice and salt were refrigerant in earlier days. Later in the year 1834, ether, ammonia, sulpher dioxide, methyl chloride and carbon dioxide were found useful as refrigerants. Air, dry ice, liquid gas (N2, O2, H2, Helium etc.) were also used later as on refrigerant. Most of refrigerant materials have been discarded for safely reasons or for lack of chemical or thermal stability. In present days many new refrigerants including halocarbon, hydrocarbon are used in refrigeration and air conditioning application. There is no ideal refrigerant, which can be used for all application. If one refrigerant has some advantage, it will have some disadvantage also. So a refrigerant which has greater advantages and less disadvantages is selected.

The refrigerant is designated as R followed by certain numbers e.g R-11, R-21, R-114 etcs. The refrigerant derived from methane base is followed by two digit, while three digit number represent ethane base.

m = Number of carbon atoms

n = Number of Hydrogen atom

p = Number of chlorine atom

q = Number of fluorine atom

in general $n+p+q = 2m+2$

So refrigerant is numbered as $R_{(m-1)(n+1)q}$

    First number on right is q (number of fluorine atom)

    Second number on right is n+1 (number of hydrogen atom plus one)

    Third number on right is m-1 (number of carbon atom minus one)

    A two digit number is actually 3 digit number with m-1 = 0 i.e. R_{011} or R_{11} simply.

1.3.1 Classification Of Refrigerant: Refrigerant is broadly classified in two groups.

(a) Primary refrigerant

(b) Secondary refrigerant

A refrigerants which directly take part in the refrigeration system called primary refrigerant, while.

A refrigerants which are first cooled by primary refrigerants and then used for cooling purpose are called secondary refrigerants.

    The primary refrigerants are further classified in four groups.

    1. Halocarbon refrigerant

    2. Azeotropic refrigerant

    3. Inorganic refrigerant

    4. Hydrocarbon refrigerant

1.3.1.1 Halocarbon Refrigerant: There are 42 halocarbons identified as refrigerant by ASHRAE (AMERICAN SOCIETY OF HEALTH, REFRIGERATION AIRCONDISINING ENGINEERS) but few are commonly used such as R_{11} ($CCl_3 F$), R_{12} ($CCl_2 F_2$), R_{13} ($CClF_3$), R_{14} (CF_4), R_{21}($CHCl_2F$), R_{22} ($CHClF_2$), R_{30} (CH_2Cl_2), R_{40} (CH_3Cl), R_{100} (C_2H_5Cl) & R_{113} (CCl_2FClF_2).

1.3.1.2 Azeotropic Refrigerant: These are stable mixture of refrigerant. These mixture usually have properties that differ from their component.

R_{500} 73.8% R_{13} and 26.2% R_{152} CCl_2F_2 / CH_3CClF_2

R_{502} 48.8% R_{22} and 51.2% R_{115} $CHClF_2$ / $CClF_2CF_2$

1.3.1.3 Inorganic Refrigerant: These were exclusively used before the introduction of Halocarbon refrigerant. These refrigerants are still in use due to their inherent thermodynamic and physical properties. Some example are.

R_{717} Ammonia NH_3

R_{729} Air

R_{744} Carbon-dioxide CO_2

R_{764} Sulpher-dioxide SO_2

R_{118} Water

1.3.1.4 Hydrocarbon: Most of hydrocarbon refrigerant are successfully used in industrial and commercial installation but these are highly inflammable and explosive e.g.

R_{170} Ethane C_2H_6

| R$_{290}$ | Propane | C$_3$H$_3$ |
| R$_{600}$ | Butane | C$_4$H$_{10}$ |

1.3.2 Type Of Refrigeration: Refrigeration is classified primarily on the basis of following two things.

(a) Refrigeration is classified on the basis of heat carrying medium in refrigeration which rejects or accepts heat i.e. refrigerant. It may be air, steam, gas or liquid etc.

(b) The refrigeration is classified on the basis of process of refrigeration i.e. compression, expansion or evaporation, absorption of a refrigerant.

The various refrigeration systems are enumerated as below.

    (a) Ice refrigeration.

    (b) Low temperature refrigeration.

    (c) Liquid gas refrigeration.

    (d) Steam jet refrigeration.

    (e) Evaporative refrigeration.

    (f) Air expansion refrigeration.

    (g) Gas throttling or (expansion) refrigeration.

    (h) Vapor absorption refrigeration.

    (i) Vapor compression refrigeration.

1.3.2.1 Ice Refrigeration: Water at 0°C reject latent heat of freezing to become ice and vice versa i.e. ice at 0°C absorbs latent heat of melting or refreezing from surrounding, lowering its temperature giving refrigeration effect e.g. Kulfi making.

1.3.2.2 Low Temperature Refrigeration: Temperature from -100°C to -273°C is in low temperature refrigeration. Low temperature refrigeration is used for solidification of CO2 and liquidification of gases i.e. air, oxygen, nitrogen, hydrogen and helium. Solid CO2 or dry ice is used as refrigerant in dry ice refrigeration. It is multistage (two or more) vapor compression or cascade refrigeration through orifices, absorbs latent heat of sublimation from its surrounding, giving refrigeration effect. **So dry ice (solid CO$_2$) is used as refrigerant in dry ice refrigeration.**

At low temperature below tripple point, solid dry ice sublimates into gas CO$_2$ and absorbs latent heat of sublimation from its surrounding giving refrigeration effect.

At low temperature above triple point liquid gas is converted into gas absorbing latent heat of vaporization from its surrounding.

To produce low temperature of gas, the gas is expanded either isentropically or by throttling from high pressure to low pressure irreversible expansion, through orifies. It gives liquidifacation of gas. As heat flows from high temperature to low temperature, so temperature of surrounding drops as it transfer heat to **liquid gas** at low temperature giving refrigeration effect.

1.3.2.3 Steam Jet Refrigeration: It is old refrigeration system in which water is used as refrigerant and its principle is that if pressure at liquid surface is reduced the boiling temperature of liquid also reduces. At normal atmospheric Pressure 76cm of Hg (1.013 bar) the water boils at 100°C, if the pressure at water surface is reduced 0.014 bar the water boils at 12°C, so if water is throttled through the jet or nozzle, it starts boiling i.e. water converts in dry vapor absorbing latent heat giving refrigeration effect.

1.3.2.4 Evaporative Refrigeration: Evaporative and air expansion refrigeration is very useful in passenger aircraft jet, fighter missiles etc. In this system air is used as refrigerant. If in expansion refrigeration, gas is used in place of air as refrigerant than system is called **Gas throttling refrigeration.**

1.3.2.5 Vapor Absorption: Vapor absorption is oldest method of refrigeration in which a gas is liquefied. In this system an absorber is used in place of compressor. Vapor from evaporator is drawn into an absorber, where it is absorbed by a weak solution of the refrigerant forming strong solution. Strong solution is heated by some external source to drive out refrigerant, which enters to condenser where refrigerant is liquidized to enter the evaporator to complete refrigerant cycle. In this system heat energy is required instead of mechanical energy of compressor.

1.3.2.6 Vapor Compression Refrigeration: This is improved type of air refrigeration, in which suitable refrigerant is used in place of air. It condenses and evaporates at temperature and pressure close to atmospheric condition. The common refrigerant are Ammonia, CO_2, SO_2 which passes in compressor evaporative condenser.

1.4 Air Conditioning:

It is an application of refrigeration system. Air conditioning is defined as the control of factor affecting the human comfort factor i.e. air movement temperature, humidity and its cleanliness or purity in required space of a building.

Air conditioning consists of both cooling and heating. So the air conditioning system should be able to add or remove heat from the desired space in addition to control air movement humidity and purity of air i.e. to remove smell dust and impurities from the air. The machine used for these comfort factor control is called air conditioner. Comfortable normal temperature and humidity of air for a person is 25°C and 50% while comfortable air velocity is 25-30 ft / min.

1.5 Air Conditioning Cycle:

Air conditioning or refrigeration works on reverse carnot cycle. On P-V diagram or 1.5 diagram it is shown blow.

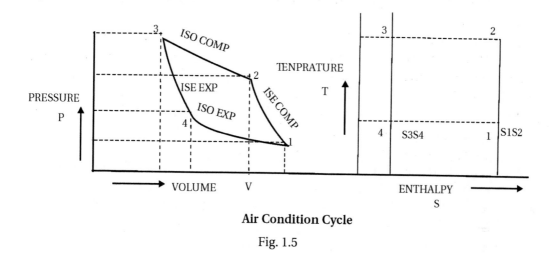

Air Condition Cycle

Fig. 1.5

The refrigerant is compressed first isentropically and then compressed isothermally i.e. at constant temperature and then refrigerant is allowed to expand first isentropically and then isothermally. During isentropic compression no heat is absorbed or rejected by refrigerant while heat is rejected by refrigerant in isothermal compression, during which its pressure increases & temperature and volume decreases. During isothermal expansion at constant temperature the pressure of refrigerant is reduced and volume increases, while in isothermal expansion, the heat is absorbed by refrigerant from cold body to increase temperature & volume of refrigerant decreasing its pressure as shown in diagram.

1.6 Heat & Moisture Transfer in Air Conditioning:

When unsaturated air comes in contact with any dry surface, then heat is transferred due to dry bulb temperature difference between air & surface and called sensible heat transfer but if saturated air comes in contact with wet surface then heat is not only transferred due to change in DBT, but also due to difference in vapor pressure of air & at wet surface. Heat transfer on account of pressure difference of vapor, results in transfer of moisture resulting condensation or evaporation which is accompanied with heat transfer as latent heat transfer. So in air conditioning total heat transfer is the sum of sensible heat transfer and latent heat transfer.

1.7 Coefficient of Performance:

Coefficient of performance (COP) of refrigeration or air conditioning machine is defined as ratio of heat extracted by machine to the work done on machine i.e. reciprocal of efficiency of machine.

Thermal C.O.P. = Q / W = 1/η η is efficiency = W / Q

Where, Q is amount of heat extracted or amount of refrigeration provided or capacity of machine.

W = Amount for work done.

For unit mass of refrigerant C.O.P = q /w

If some heat is lost or goes waist then C.O.P = $\dfrac{\text{Actual COP}}{\text{Theoretical COP}}$

1.8 Diffrence Between A Heat Engine, Refrigerator And Heat Pump:

An heat engine is driven in which heat is supplied from hot body to cold body, while refrigerator is reverse of heat engine in which heat is extracted from cold body and supplied to hot body. In both heat engine & the refrigerator temperature of cold body is below the atmospheric temperature.

If Q_2 is heat supplied by engine & Q_1 is heat rejected by engine than work done by engine.

$W_2 = Q_2-Q_1$

Then COP or efficiency of engine $(COP)_E$ or $\eta_E = W_E/ Q_2 = \dfrac{\text{Work done}}{\text{Heat supplied}} = \dfrac{Q_2-Q_1}{Q_2}$

In case of refrigerator $(COP)_R = 1/ \eta_R = \dfrac{\text{Heat extracted}}{\text{Work done}} = \dfrac{Q_1}{Q_2-Q_1}$

In heat pump heat is supplied from cold body to hot body by doing some work by pump as in refrigerator. The temperature of cold body is above atmospheric temperature as in case of heat engine but in case of refrigeration the temperature of cold body is less then the atmospheric temperature schematically they are shown below.

$(COP)P$ = Energy performance ratio E.P.R = $Q_2 / W_P = Q_2 / Q_2 - Q_1$

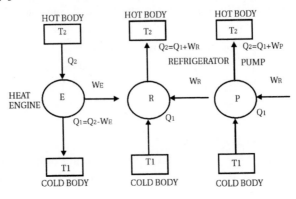

Fig. 1.8

1.9 Examples:

Example (1): Find COP of a refrigeration system if the working input 180 KJ / Kg and refrigeration effect produced is 200 KJ / Kg.

Solution: Given for unit mass q = 200 KJ / Kg

& W = 100 KJ

COP of refrigeration system = 200 / 100 =2

Example (2): A refrigeration machine work between 300°K and 240°K determine COP.

Solution : Given T_2 = 300°K T_1 = 240°K

$(COP)_R$ = T_1 / T_2 - T_1 = 240 / 300 - 240 = 240 / 60 = 4

Example (3): Find the coefficient of performance of machine working between 310°K & 260°K and operating as (a) Heat pump (b) Heat Engine (c) Refrigerator.

Solution: Given T_2 = 310°K T_1 = 260°K

(a) COP of heat engine = (T_2-T_1) / T_2 = (310 - 260) / 260

= 50 / 310

= 0.16

(b) COP of heat pump = T_2 / (T_2-T_1) = 310 / (310 - 260)

= 310 / 50

= 6.2

(c) COP of Refrigerator= T_1 / (T_2-T_1) = 260 / (310 - 260)

= 260 / 50

= 5.2

Also $(COP)_P$ = 1 / $(COP)_E$ = 1 / 0.16 = 6.2

$(COP)_R$ = $(COP)_P$-1 = 602 -1 = 5.2

Example (4): A cold storage to be maintained at -5°C while the surrounding air temperature is 30°C. The heat leakage from the surrounding into the cold storage is estimated to be 24KW. The actual COP of the refrigeration plant is 1/3 of an ideal plant working between same temperature find powers to drive the plant.

Solution: Given T_1 = -5°C   = (-5 + 273°K) = 268°K

T_2 = 30°C   = (30+ 273°K) = 303°K

Heat leakage from surrounding Q_1 = 24 KW, (COP) actual = 1/3 (COP) Ideal

(COP) ideal   = T_1/T_2 - T_1 = 268/303 - 268 = 268/35 = 7.6

(COP) actual = 1/3 (COP) idea = 7.6 /3   =2.86

Since (COP) = Q_1 / W_R W_R = Q_1 /(COP)

Since temperature is to be maintained so heat extracted must the heat leakage from the surrounding.

W_R = 24 / 2.86 = 8.39 KW

Example (5): 1.5 KW per tone of refrigeration (TR) is required to maintain the temperature of -40°C in the refrigeration. If the refrigeration cycle works as carnot cycle, determine the following.

(1) COP of cycle (2) Temperature of sink (3) Heat rejected to the sink per tone of refrigeration (4) Heat supplied and EPR if cycle is used as a heat pump.

Solution: Given T_1 = -40°C + 273 = 233°K

1.5 KW power per TR is required i.e. W_R = 1.5 KW

= 1.5 KJ/s

= 1.5 / 1 /60

= 90 KJ /m

Heat extracted by refrigerant Q1 = ITR

= 210 KJ /m

(i) (COP) = Q_1 / W_R = 210 / 90 = 2.33 Ans.

(ii) Let Te = temperature of sink (hot body)

$(COP)_R$ = T_1 / T_2 - T_1 = 2.33 so T_1 = (T_2 - T_1) 2.33

Or (1 + 2.33) T_1 = 2.33 T_2

Or T_2 = 3.33 / 2.33 x 233 = 333°K =333 - 273 = 60 °C

(iii) Head refectedQ_2

Q_2-Q_1= $\dfrac{Q_1}{(COP)_R}$ = $\dfrac{1}{2.33}$

Q_2= 1+ $\dfrac{1}{2.33}$ = 1.4$_T$

Q2 = 1.4 Ton

(iv) For Heat Engine

EPR = Q_2/Q_2 - Q_1

= T_2 / T_2 -T_1

= 333 / 333 - 233 = 3.33 Ans

Q_1 = 210, EPR= 3.33

So Q_2/Q_2 -210 = 3.33   Solving it Q_2 = 300 KJ / m Ans.

Example (6): Six hundred Kg of fruit are to be kept in cold storage for preservation temperature atmospheric temperature is 20°C. The cold storage is to be maintained -5°C and fruits get cooled to storage temperature in 10 hrs. The latent heat of freezing is 105 KJ / Kg and specific heat of fruit is 1.256 KJ / KG K . find refrigeration capacity of plant.

Solution: Given T_1 = -5°C = -5 + 273 = 268

$T_2 = 20°C = 20 + 273 = 293$

Mass of fruit is $= 600$ Kg $C_F = 1.256$

Sensible heat to be removed or sensible heat load	$= m \, C_F \, (T2 - T1)$
	$= 600 \times 1.256 \, (293-268)$
	$= 600 \times 1.256 \times 25$ KJ
	$= 15000 \times 1.256$
	$= 18840$ KJ
Latent heat of freezing or latent heat load	$= mL_F$
	$= 600 \times 105$
	$= 63000$ KJ
Total heat load or heat to be removed	$= 840 + 63000$
Q in 10 hrs.	$= 81840$ KJ
Q per min	$= 81840 \, / \, 10 \times 60$
	$= 136.4$ KJ /m
Refrigeration capacity of plant to tone	$= 136.4 \, / \, 210$
	$= 0.64$ TR Ans.

Example(7): A cold storage plant is required to store 10 tones of fish. Fish is supplied to cold storage for preservation at atmospheric temperature 30°C. The Sp. heat of fish above freezing point is 2.93 KJ / Kg K and Sp heat of fish below freezing point is 1.26 KJ /Kg K. The cold storage is maintained at -8°C and freezing point of fish -4°C, latent heat of fish is 235 KJ/ Kg power required to drive ac plant is 75 kw. Find (I) The capacity of ac plant (II) Time taking to achieve cooling (COP) actual is 1/3 of the (COP) carnot.

Solution: Given Mass of fish m = 10 tones = 10 x 1000 = 10,000 Kg

T_2 temperature of hot body (atmospheric) = 30°C = 30 + 273 = 303°K

T_1 temperature of cold body (cold storage)= -8°C = -8 + 273 = 265°K

Freezing point of fish T_3 = -4°C = -4 + 273 = 269°K

Latent heat of fish LF = 235 KJ / Kg

Power W_R in KW = 75 KW = 75 KJ / sec = 75 x 60 = 4500 KJ /m

1 Capacity of the plant

(COP) ideal or carnot	$= T_1 \, / \, T_2 - T_1 = 265 \, / \, 303 - 265 = 6.97$
W_R	$= 4500$ KJ / m
Capacity of plant $(COP)_R$	$= 1/3(COP)$ ideal or carnot $= 6.97 \times 1/3 = 2.03$
So heat removed from plant Q	$= (COP)_{actual} \times W_R$
	$= 2.03 \times 4500$ KJ/m
	$= 9408$ KJ/min
So capacity of plant	$= 9408/210 = 44.8$ TR Ans.

2. Time required:-

Sensible heat load of fish above freezing point $= m\, C_{FISH}\, (T2 - T3)$

$= 10{,}000 \times 2.93\, (303 - 269)$

$= 1.99 \times 10KJ$

Sensible heat load of fish below frizzing point $= m\, C_{FISH}\, (T3 - T1)$

$= 10{,}000 \times 1.26\, (269 - 265)$

$= 0.101 \times 10KJ$

Latent heat load $= m\, L_{FISH}$

$= 10{,}000 \times 235$

$= 4.7 \times 10KJ$

Total heat removed $= (1.992 + 0.101 + 4.7)\, 10KJ$

$= 6.793 \times 10$

Time required to achieve cooling in hrs $=$ Total heat removed/heat per hrs.

$= 6.793 \times 10\, /\, 9408 \times 60$ hr.

$= 67.93\, /\, 5.6448$

$= 12$ hr.

1.10 Unit of Air Conditioner or Refrigeration Plant:

Refrigeration unit is tone of refrigerator which is defined as capacity or latent heat required to convert a ton (1000 Kg or 2000 Ibs) of Water at 32°For 0°C into ice in 24 hr.

Latent heat of freezing $= 80$ Kcal / Kg (144 Btu/1b)

One ton of water converted in 0.9 ton of ice

Ton of refrigeration ML $= 900 \times 80 = 72000$ Kcal / day

$= 72000/24 = 3000$ Kcal / hr

$= 3000/60 = 50$ K cal /m

1 TR $= 3000$ Kcal / hr

$= 50$ Kcal/min

$= 210$ KJ / min

$= 3.5$ KJ / sec or 35 KW

One tone of refrigerator is also defined as production of cold at the rate at which heat is to be removed from one ton of water 0°C to freeze to be ice at 0°C in one day.

1.11 Psychrometry Of Air Conditioning Process:

The branch which deals with alteration state of moist air in any air conditioning process. In air conditioning normally mixing of two kind of air conditioning at different temperature & humidity takes place.

1.12 Mixing Process In Air Conditioning:

In adiabatic process of mixing of two state of dry air mass ma, ma2, H ma3 the mass of mixer i.e. $m_{a3} = m_{a1} + m_{a2}$

The equation of enthalpy = $m_{a3} h_3 = m_{a1} h_1 + m_{a2} h_2$

i.e. Enthalpy of mixer will be between enthalpy of two states. Psychrometry chart between specific heat (ω) and enthalpy (h) is straight.

The position of mixture state divides the w-h line in reverse ration of masses of dry air in two state in this mixture process, the heat transfer or gain or loss is became of sensible heat transfer and latent heat transfer.

$$\text{Sensible heat factor SHF} = \frac{Q_S}{Q_S + Q_L}$$

$$= Q_S / Q$$

$$= \frac{\text{Sensible heat}}{\text{Total heat}}$$

$$= \frac{(h_A - h_B)}{(h_C - h_A)}$$

Sensible heat process $Q_S = m_a (h_B - h_c)$

$$= m_a (t_B - t_C)$$

Latent heat process $Q_L = m_a h (\omega_C - \omega_A)$

1.13 The Components of Refrigeration or Air Conditioning Unit:

The basic components of refrigeration system or air conditioner are (a) Evaporator (b) Compressor (c) Condenser.

1.13.1 Evaporator: It absorbs heat from surrounding. The evaporator is made of copper tube coil through which refrigerant flows. In air conditioning system, the evaporator coil is located in air duct so that it removes heat from the air as soon as it is made to pass over it at some velocity. The evaporators coil also removes moisture from air in air conditioning process.

1.13.2 Compressor: A compressor is main component of heat vapor compression refrigeration and air conditioning system. The main function of compressor is to raise pressure of the refrigerant and pump the refrigerant at high pressure into condenser. The compressor is driven by an electric motor and during the operation of compressor, some heat is added to refrigerant and some of which is rejected into the lubricating oil or any other cooling medium. The compressor and condenser both heat rejecting component are located near each other.

1.13.3 Condenser: The main function of the condenser is to reject heat absorbed by evaporator, which is made in an coil shape with large plate area with fins or without fins to remove heat by natural air circulation in case of air cooled condenser. Often a fan is used to cool the condenser. For larger capacity of condenser in big air conditioning plant, water is used to cool refrigerant, which is circulated around the refrigerant to absorb its heat and its vapor is condensed. In case of plant of small refrigeration capacity only air cooled condenser is located inside building or space e.g. domestic refrigerator, but in larger refrigerant plant, condenser located outside building not to add excess heat in the building to be cooled e.g. the cold storage or ice plant or chiller air conditioning plant, water cooled condenser are located outside the building.

Beside above three components we use refrigerant control valve or metering device also called as expansion valve and thermostatic switch.

1.13.4 Refrigerant Control Valve: It controls amount of refrigerant to enter into evaporator. There are three type of control valve in refrigeration or air conditioner system namely capillary tube, automatic expansion valve or thermostatic expansive value. A capillary tube is a copper tube of approximately 3cm dia and 30cm length. The exact dia and length are found by calculating amount of refrigerant required to enter evaporator. We may use some times two capillary tubes.

1.13.5 Air Outlet : Finally cooled & dehumidified filtered air enter in the room or space to be air conditioned through air outlet properly placed.

They may be of following type.

(i)Supply outlet (ii) Ceiling diffuser outlet (iii) Grill or vane outlet (iv) Register outlet.

1.13.6 Thermostatic Switch / Control: Its main function is to sense the temperature of the substance (or the surrounding) required to be cooled and switch on or off automatically the air conditioning unit as per requirement of cooling to keep the substance of surrounding at a particular temperature on thermostatic control.

1.14 Equipment used in an Air Conditioning System:

The various components used in air conditioning system are as below.

1.14.1 Air Conditioner Unit: It consist of component machine for cooling and dehumidifying air for summer air conditioning or heating or humidification for winter air conditioning.

1.14.2 Circulation Fan or Air Handling Unit: Its main function to control movement of air to & from the room to be air conditioned.

1.14.3 Supply Duct: It directs the conditioned air from the circulating fan to the space to be air conditioned at proper point, we shall discuss it later on in detail.

1.14.4 Supply Outlet: These are grills which distributes the conditioned air evenly into room.

1.14.5 Return Outlet: These are opening in a room which allow the room air to enter the return duct.

1.14.6 Filters: Its main function is to remove dust, dirty fragrances (bed smell) and other harmful bacteria's from air i.e. to purify air.

1.15 Classification of Air Conditioning:

The air conditioning system may be broadly classified in three ways.

1.15.1 According to Purpose or Use It Can Classified As:-

(i) Comfort air conditioner

(ii) Industrial air conditioner

1.15.1.1 Comfort Air Conditioner: The air is brought to be at 21°C dry bulb temperature and 50% relative humidity for the human health, comfort and efficiency and sensible heat factor is generally kept fro different building as following.

For residences or private office   = 0.9
For restaurant and busy offices   = 0.8
Auditoriums or cinema hall   = 0.7
Ball room dance, hall etc.   = 0.6

This air conditioning is adopted for home, offices, shops, restaurants, theaters, hospitals, schools etc.

1.15.1.2 Industrial Air Conditioning System: It is important system now - a - days, in which inside dry bulb temperature and relative humidity of the air is kept constant for proper function of the sophisticated machine and proper rosercl and manufacturing process. The system is used in textile or paper mills, various machine part manufacturing plant, tools room, photo processing plants etcs.

1.15.2 According to season of the year:

According to season of the year the air conditioning is classified as

(i) Winter air conditioning system

(ii) Summer air conditioning system

(iii)Year round air conditioning system

1.15.2.1 Winter Air Conditioning System: In this air conditioner the air is heated followed by humidification the schematic arrangement of the system is shown in figure.

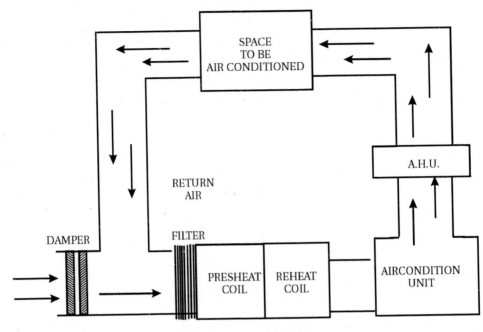

Winter Air Conditioning System
Fig. 1.15.2.1

The out side air flows through a damper and mixed up with the returned circulated air obtained from the conditioned space. The mixed air passes through a filter to remove dirty dust and other impurities. The air is then passed through a preheat coil in order to prevent possible freezing of water and to control the evaporation of water in humidifier. The preheated air is passed through some reheat coil to bring the air to desired / designed dry bulb temperature. The conditioned air is supplied to the space to be conditioned by a blower fan or air handling unit, a part of the used air is exhausted to the atmosphere by exhaust fan or ventilator. The remaining part of the used air (returned circulated air) is again conditioned as shown in figure. Such systems are central heating plant. These are not available in window/package type air conditioner. But heating element/coils are fitted in summer window type air conditioner to give hot air in winter heating element, steam or hot water in surrounding pipe is used to heat the air in central heating plant.

1.15.2.2 Summer Air Conditioning System: It is most important and prevalent air conditioning system in which air is cooled and dehumidified. The schematic arrangement of summer air conditioning system is shown in figure.

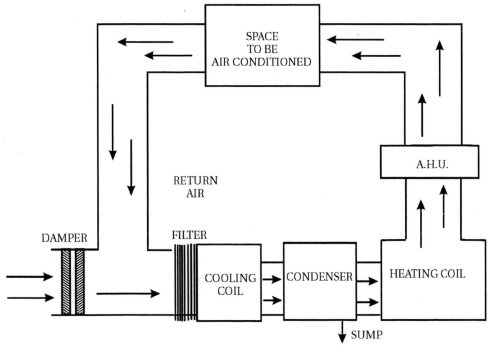

Summer Air Conditioning System
Fig. 1.15.2.2

The out side air is sucked through damper and mixed with return air coming from the conditioned space. The mixed air, after passing through a filter to remove dust, dirt and other impurities, is passed through cooling coil. The cooling coil has temperature much lower, then required dry bulb temperature. The cooled air passes through perforated membrane condenser to loose its moisture in condensed form, which is collected in sump. After that this air is passed through a heating coil or steam from boiler which heats up the air slightly to bring designed dry bulb temperature and relative humidity. The air thus conditioned is passed to the space to be air conditioned via AHU or blower in which part of air is exhausted to the atmosphere and remaining air is recirculited as discussed earlier. Such air conditioners are used for cooling only. They are available in window/package type or center air conditioning.

1.15.2.3 Year Round Airconditioning System: This modern system should have both the summer and winter air conditioning. In summer cooling coil work while in winter cooling becomes inoperative. The dehumidification is done in summer by operating cooling coil at a temperature lower then the dew point temperature, while in winter humidification is made to use with heating coil. Such plants are central air conditioning plant. Now a days conditioned air is obtained by operating refrigerant by reversing cycle. Such air conditioners are available both as window / split / package type unitary air conditioner and central air conditioner. They are called heating cooling room or unitary air conditioner or central air conditioner.

1.15.3 According To Arrangement of Equipment:

We can classify air conditioning system according to arrangement of equipment.

a. Unitary air conditioning system

b. Central air conditioning system

1.15.3.1Unitary Air Conditioning System: In this system factory assembled air conditioner are installed in or adjacent to the space to be conditioned. These systems are of following three kind.

1. Window Air Conditioner: These are self contained unit of small capacity 1 TR to 4 TR and are mounted in a window or through the wall. They condition the air of one room only. If the room is bigger in size, the two or more units are installed.

2. Vertical Packed Unit: These are also self contained units of higher capacity 5 to 20 TR and installed adjacent to the space to be air conditioned. The conditioned air is supplied to the space via insulated pipe or duct. This is very useful for air conditioning the air of a restaurant, bank or small office or two, three rooms of bigger office.

3. Split: These air conditioners are split in two units indoor & outdoor connected through a pipe. The indoor unit is kept in side room, while outdoor unit can be kept anywhere outside. These are useful where window type can be installed to get rid of exhaust hot or cooled air. This unit of air conditioner system may be adopted for winter, summer or year round. The air of the unit exhausted by conditioner in the vicinity.

1.15.3.1.1 Window or Package Air Conditioner: It consist of following components in a steel box as shown in figure. Window air conditioners are installed near to the floor or at window height or in window.

(A) Compressor: Compressor of suitable capacity is driven by electric motor. It may be centrifugal, rotary or screw type. It is situated in middle between condenser and evaporator as shown in the figure.

(B) Condenser: Water cooled or air cooled condenser which uses cooling fins or propeller fan to discharge the heat of condenser, refrigerant to out side atmosphere. It is situated in rear side as shown in figure.

(C) Evaporator: Expansion type cooling copper tube with thin fins is situated near outlet grill as shown in figure.

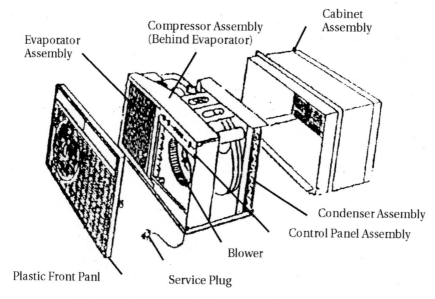

Fig. 1.15.3.1.1 (A) Window Air-Conditioner

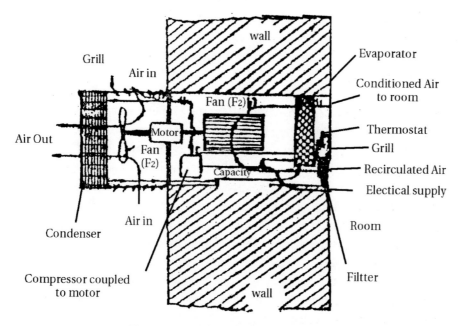

Fig. 1.15.3.1.1 (B) Window Air-Conditioner

(D) Expansion Valve: Thermostatic expansion valve is used.

(E) Refrigeration Line: Copper pipes are used to carry refrigerant.

(F) Drier Unit: It is used to absorb moisture and other contaminant in expansion valve.

(G) Filters: These are used to purify the air.

(H) Blower: Condensing evaporating line cools the air, which can be sent to the room by blower fan and damper control when circulate fresh filtered & cooled air, after exhausting smoky polluted air inside.

Window air conditioners are installed near to the floor or at window height or in window.

1.15.3.1.2 Split Air Conditioner: In split air conditioner, we have two unit, indoor, outdoor unit i.e. unit is split in two units. The indoor unit is lighter and with grill having weight 12 & 14 Kg approximately 1/5 or 1/10 of the weight of outer unit. The indoor unit consists of blower fan, and a reverse cycle valve, if it is a heating cooling air conditioner. The outer unit has compressor & its electric motor, condenser, evaporator & filter. The air circulation system in summer split air conditioner is such that the cooled air blows into room & polluted and warm air is sucked and exhausted outside. Both unitary split window or package air conditioner can work on reverse cycle system i.e. they can be used for both winter or summer. For winter air conditioner cycle is reversed i.e. warm air blown inside & cold air is sucked. This is done by changing the direction of valve in indoor unit and capacity of these air conditioner are selected as per heat load. Fresh and cooled air from evaporators of package units is sent to adjacent rooms via duct.

1.15.3.2 Central Air Conditioner: This is most important type of air conditioning system which is adopted for big building, when cooling or heating capacity (heat load) required is 25 TR or more and air flow required is more than 300m³ per minute or different zone in a building is to be air conditioned, as the cost of many window / package unitary air conditioner units will be more than the cost of central air conditioning system. In this system compressor, condenser and evaporator are placed out side building. The specification and capacity of the air conditioning system is selected as per heat load, while air handling unit is designed as per dehumidified air quantity.

Fresh and cooled air from evaporator is sent to room or other part of the building by air handling unit AHU via duct specially designed as per level, air speed and friction loss and heat leakage i.e. length of short or big arm of duct are designed as per sound proper aspect ratio. In winter air is blown by air blower damper control or AHU through heating strips or furnace in normal air conditioning, while compressor and condenser etc. i.e. refrigeration unit is put off. Central heating plant are installed on high hills, where cooling is not required. These Plants use steam from a boiler to give hot air which is blown to building through duct and proper AHU. All walls, windows and doors shutters, ceilings are made well insulated and air tight to avoid heat leakage.

In modern central air conditioning VRV system, working on reverse cycle as heating cooling window or split air conditioners are very common in application now-a-days. No separate boilers, heating strips or fins are required to heat air.

The most common central air conditioning (cooling) plant is chiller air conditioner.

1.16 Chilling Machine Or Chiller Air Conditioner:

The water chilling machine consists of following components.

1.16.1 Scroll Compressor: The unit has a scroll open type or bolted hermetic compressor operating with R22 refrigerant and speed of 3500 rpm to 50 Hz. The compressor housing is made of high grade cast iron precisely machined to provide a very close to clearance between rotor and housing. The rotor is mounted on multiple cylindrical or tapered roller bearings with built in reservoir of lubricant oil for lubrication of all bearings and check valve to prevent back spin during shut down. An oil pump with an oil heater is provided for circulation of oil lubricant. It has a shaft seal to prevent leakage of refrigerant. An oil separator with suitable heat exchanger is also provided to isolate oil with refrigerant. The driving motor for rotor is double squirrel cage or suitable hermetic type as required with protection to avoid damage.

1.16.2 Condenser: The circulated condenser coils have aluminum fins bounded to seamless copper tube and are in minimum three rows deep with at least 12 fins per inch. The number of direct driven fans with a safety guards and with low noise level are arranged for horizontal or vertical discharge of air. All controls like fuse selector switch, oil safety switch, high and low pressure cut outs, crankcase heater interlocks and various protection devices are housed in a weather proof enclosure.

1.16.3 Chiller Evaporator: There are two types of chiller evaporators' direct expansion or dry expansion (Dx) type. The refrigerant flows in seamless copper tubes or pipe cooler, supported in a chamber or shell by adequate stiff support to minimize vibration and noise in which cooling water or medium flows. These shall be constructed with steel by welding and fitted with steel sheets on either side. The baffles are provided in shell to ensure adequate water velocity in shell over tube. The refrigerant header is made of cast iron and tubes are expanded at ends in header to prevent leakage. The cooler is designed for a pressure of 21 Kg/sq cm tested with N2 or CO2 gas against leakage and is insulated with expanded polystyrene or equivalent insulation, while chiller shell is tested at 9 Kg/sq cm. The chiller shells and tubes may be multi-pass or multi-circuit and provided with thermostatic expansion valves, control solenoid valves, drier filler relief valve with connected drain valve, thermo meter and pressure gauge at both water inlet and outlet.

1.16.4 Flooded Type Evaporator: The out side shell and chamber, made of steel by welding with steel sheets on either side, surround chiller pipe in witch water or cooling medium flows while the refrigerant flows in shell i.e. flooded in shell around chiller tube. Tube ends are expanded in tube sheets to avoid leakage of refrigerant and the tubes are adequately supported at 1m interval. Refrigerant shell side is tested at 2Kg/sq cm and chiller tube is designed at 14Kg/sq cm and tested at 8 Kg/sq cm. This type of chiller has moisture indicators and side class with refrigerant filter drier, with two stage hot gas muffler, thermostatic expansion valve, control solenoid valve, changing valve. Suction lines are insulated with polystyrene, ethane insulation and aluminum cladding.

Chilled water compressor and cooling tower is selected on the basis of ton capacity of plant, while condenser or chiller evaporator and its outlet is decided on the plant capacity i.e. compressor and cooling tower and AHU is designed on the quantity of dehumidified air. The specifications of condenser, chilling pump and piping are selected as per need of water.

1.16.5 Water Pump Set: A primary water centrifugal pump with end suction vertical casing with a 3 phase, 415 V, TEFC induction motor and secondary water variable speed pump for various zone are used. Each zone has 3 number 415V, 3 phase, 50 Hz and section vertical split casing induction motor and one no. micro processed based pump logic controller.

1.16.6 Motor and Starter: Three phase 50 Hz 415 volts AC supply is needed for TEFC squirrel cage induction motor designed and guaranteed for continuous operation. The motor is provided with protection against operation in excess of rated supply. Starters are star-delta winding type.

1.16.7 Control Console: Chiller unit has microprocessor based control console for outdoor unit with all operating and safely controls with display arrangement for water temperature, evaporators pressure, condenser pressure, system voltage, compressor ampere run time for each compressor or indicating number of compressors working, time relay for protection of compressor, self and auto start run and stop arrangement.

Apart of these above components chiller unit has coupling guard or belt to couple motor & condenser. Steel frame and RCC foundation are required to mount these components with vibration pads. A dust proof sheet steel control panel is also provided with unit for housing controls, gauges, thermostatic switch etc. with LED light indicators for safety device. Panel also has a push button and a indicating lamp for continuity testing of heating element.

On installation on foundation refrigerant circuit and water chilling pipe and connecting parts are checked for leakage. Power consumption and capacity in tons is also checked by current of motor and measurement of flow rates of water and temperature difference. Alignment of motor and compressor should also we checked before starting of unit.

1.16.8 Air Handling Unit: This unit is floor mounted or ceiling suspended belt driven or direct driven fan with 3 phase 415V motor. The units are of different capacity at static pressure for requirement of air conditioning of space. It is double skin modular construction of fan section, coil section, filter section, insulated drum pan etc. The units are factory fabricated with preplastified GSS heat and polythene foam insulation. The die cast aluminum panels are fixed with handle, hinges and latches. Heating or cooling coils have aluminum fins mechanical bonded to coppers, with proper silver brazed header and bends of proper capacity and with 2 way modulating valve and actuator, control wiring & thermostat etc. The units are also provided with pre-filter sets for fresh air intake, drain connection with thermometer and pressure gauges at both inlet and outlet. AHU are mounted on concrete or steel foundation. Aluminum fresh air louvers, dampers are also provided with AHU at fresh air intake.

1.16.9 Duct: The conditioned air (cooled or heated) from the air conditioner equipment must be properly distributed to room or space to be air conditioned to provide comfort condition. When conditioned air cannot be supplied directly to room or space, then ducts are used, which convey the conditioned air to proper air distribution point or air supply outlet in rooms and carry the return air from the rooms to air conditioning equipment for recirculation after reconditioning.

Duct system cost nearly 20 to 30% of the total cost of air conditioned plant and power required by fan of AHU forms a substantial part of the running cost(energy consumption etc;), so air duct system is so designed so that both capital cost & running cost is lowest.

1.16.9.1 Classification of Duct: The duct may be classified as per kind of carrying air, its pressure and velocity and shape of duct.

(a) Supply Air Duct: The duct which supplies the conditioned air from conditioning equipment to the room is called supply air duct.

(b) Return Air Duct: The duct which carry air from conditioned room back to conditioning equipment is called return air duct.

(c) Fresh Air Duct: The duct which carry out side air is called fresh air duct.

(d) Low, Medium and High Pressure Duct: When the static pressure in the duct is less than 50 mm of water gauge. 50mm to 150mm of water gauge and from 150mm to 250mm of water gauge, are called low, medium and high pressure duct respectively.

(e) Low & High Velocity Duct: When the velocity of air duct is up to 600m/min or more than 600m/min the ducts are referred as low or high velocity duct respectively.

The ducts can be also classified as per shape and layout of the ducts as

(1) Circular, rectangular & square duct system

(2) Loop or radial parameter duct system

(3) External plenum duct system

1.16.9.2 Duct Material & Construction: The ducts are usually made of G.I sheet, aluminum sheet or black steel sheet, among them 26 to 16 gauge galvanized iron sheet is most common because of zinc coating preventing rusting to avoid cost of painting. The aluminum sheet is used because of its lighter weight and resistance to moisture. To with stand high pressure black steel sheet is used. Now a days resin bonded glass fiber non metallic ducts are used for low velocity and low static pressure application. It can be manufactured to desired shape easily and are quite strong. The cement asbestos duct may also be used for under ground air distribution.

The sheet metal ducts expand and contract as they heat and cool, so fabric joints are often used to absorb this movement and prevent noise of fan and furnace. In most of the ducts, joints are made of sheet metal as duct.

The joints must be air tight and strong, so many joints are riveted to add strength and sealed with special duct tape to make leak proof or sir tight. When duct passes thorough unconditioned space they are insulated to reduce heat loss or gain and noise. The insulation is fastened with adhesive or metal clip in some case of larger duct under high pressure. The reinforcement prevents bulging or collasing of the duct.

1.16.9.3 Duct Shape: The duct may be circular, rectangular or square. Circular ducts are preferred as it carry more air in less space, requires less duct material due to less duct surface and so it has less surface friction and less insulation is needed in it. By good appearance point of view rectangular or more practical square ducts are used, as it present flat surface for easier to work.

Continuity equation and Bernauli's equation are very useful for design of duct. Friction dynamic pressure loss at opening and joint of duct are also to be taken into account in design of shape & size.

For lay out of the duct w.r.t to the conditioning equipment for plant and air outlet many systems are adopted namely, duct system, loop parameter, radial parameter system or extended plenum system.

1.16.9.4 Design Of Duct: The design of duct, depends on the available space, where the duct can be laid, air friction loss, sound level, heat loss or gain, heat leakage, air velocity etc.

Air velocity is taken as per comfort air conditioning of various building per following.

(a) For commercial building

Low velocity	1200 - 2500 ft/m
High velocity	above 2500 ft/m

(b) For Factory

Low velocity	2200 ft/m

For other building

As per sound level   As per duct friction

Type of building	Control factor	Control factor			
	Main duct	Main duct		Branch duct	
	ft/m	Supply	Return	Supply	Return
Residential building	500	1000	800	-	-
Hospital building	1000	1500	1300	1200	100
Direct room/library	1200	2000	1500	1600	1200
Auditorium	800	1300	1100	1000	800
Ordinary office	1500	2000	1500	1600	1200
Industrial place	2500	3000	1800	2000	1500

Selecting proper air velocity, the cross section area of duct or size of duct is calculated as follows.

$$\text{Area of cross section A} = \frac{\text{Quantity of air (Q)}}{\text{Velocity of air (V)}}$$

Knowing the area of cross section the length & height of a rectangular or diameter of circular duct can be found by design table provided by manufacturer Carrier or Voltas for minimum aspect ratio.

1.16.9.5 Aspect ratio:- The ratio of long arm to the short arm of duct is called aspect ratio.

1.16.10 Air Outlet: Finally conditioned air enter in the room or space to be air conditioned through air outlet properly placed. They are of following type.

(a) Supply or suction outlet

(b) Ceiling diffuser outlet

(c) Horizontal / Vertical vane or grill

(d) Registers

(e) Cassette type (one way or four way)

(f) ducted type (low, medium, high, static)

(g) ceiling, floor, wall mounted.

1.16.10 Design of Air Outlet: It is decided as per air outlet or terminal velocity. The various terminal velocities for different buildings taken as mentioned below.

Residence	300	-	500 ft / min
Sound controlled office	500	-	750 ft / min
Ordinary office	1000	-	1250 ft / min
Cinema hall			1000 ft / min
Hotel bed room	500	-	700 ft / min

Size or number of air outlet / grill can be found as per design. The blow of air by air outlet must be such as to cover 3 distance of air outlet to front wall. The reach of air is called blow.

1.16.11 Wiring: A suitable size of cable as per design with respect to capacity of plant and its components such as various motor and pump set are used to feed power supply to operate compressor, condenser, and other units of chiller plants.

1.16.12 Advantage of DX Type Chiller: The dry expansion type chiller evaporator has following advantage so it is widely used.

(a) Low initial temperature

(b) Low requirement of refrigerant

(c) Low possibility of freezing up of refrigerant

(d) Low mechanical problem

(e) Requires least space i.e. compact unit

1.17 Digital Scroll Compressor:

The axial compliant sealing technology adjusts the axial moving range of stator scroll pan. It has an additional by pass between section inlet and the pressure bore at the floating sealing point of the axial stator.

When the pulse width modulation (PWM) solenoid valve is closed, the two stators get engaged to achieve loading but when PWM solenoid valve of by pass is opened, the high pressure bore and low pressure inlet get connected, higher pressure in compressor bore moves the compressor a little upward resulting in unloading of compressor.

Periodical loading and unloading of the scroll compressor control the refrigerant volume. One control cycle, consist of one unloading and one loading takes 10 to 20 seconds. The control of ratio of unloading and loading times can achieve different out puts of the refrigerant e.g. the full capacity of compressor is 8 ton and control cycle is 20 seconds, if the out put is 1.6 TR (20% of full capacity) then loading time will be 20% of total time 20 second i.e. 4 second & unloading time 16 seconds. Similarly for other required capacity or load loading and unloading time can be worked out. So compressor works on variable refrigerant volume and energy is saved.

1.17.1 Advantage Of Digital Scroll System Over Inverter System:

	Inverter system	Digital scroll system
(1) High efficiency	Lower EER	Higher EER and higher COP (Energy efficiency ration) & lower COP (Coefficient of operation)
(2) Power saving	Because of 10% conversion loss and due to variable frequency and voltage	because low conversion loss lesser start ups and shut down higher EER power consumption

	control energy losses are high.	is less. So energy is saved about 33.3%.
(3) Better range	System operates between 30% to 100%	Systems operates between 10% to of capacity 100%
(4) Compressor capacity adjustment	System change between 30% to 1005 in 27 step, so adjustment tales much time	Steps modulation from 10% to 100% capacity, so response to to change in indoor conduction
(5) Stability & safety	Oil return system is needed to bring oil back to compressor during full capacity period. While during low capacity period lower speed mark oil return is efficient.	refrigerant velocity is sufficient to bring oil back to compressor during high capacity period. No problem during low capacity, oil does not leave compressor during unload period.

1.18 VRF Air Conditioner:

VRV or VRF air conditioner system consist of two main unit (a) out door unit (b) Indoor unit.

1.18.1 Outdoor Unit: This consists of rotary / scroll modular digital type compressor. Manufacturing companies manufacture different outdoor units of different capacity with their different brand name. Carrier VRV modular system is available in various capacity module, most common module is of 48 HP, while Voltas maintaining 5 basic module of MDVM multi digital variable 8HP, 10HP, 12HP, 14HP, 16HP which can be freely assembled to give desired capacity plant. Similarly Daikin is manufacturing VRV VIII model. Both carrier and Daikin manufactures 5HP, 8-10HP, 12-14-16-18 HP, 20-22-24-26 HP, 30-32-34-36 HP, 38-40-42-44-46 HP, 48-50-52-54HP. These units consist of high capacity compressor, heat exchanger condenser, fan, fan motor sine wave inverter, heat transfer piping circuit and compact aero box. Scroll compressor minimizes heat losses to give energy saving. Heat exchanger & compressor are such to give high compression. The sine wave inverter gives smooth rotation of DC motor. By performing super cooling before expansion, the volume of refrigerant that flows to the indoor unit can be reduced as per requirement of indoor unit without lowering evaporation temperature.

1.18.2 Indoor Unit: It consist of a DC fan motor to operate quiet shear fan to blow air in the room or space to be air conditioned. The incoming air is cooled by giving heat to incoming refrigerant from outdoor unit. A drain pump is also equipped in indoor unit with proper lift. The indoor unit may be cassette or ductable type, built in or suspended type, wall mounted or floor standing type or ceiling mounted type.

1.18.3 Controller: VRV system has wired or wireless remote controllers with a master controller to control individual indoor unit or it has centralized control system which can control all indoor units. These controllers has ON / OFF facility to start / stop outdoor unit & indoor unit i.e. the complete plant Including outdoor unit can be closed or shorted copper piping with proper insulation run from outdoor to indoor unit.

1.18.4 Copper Piping: The copper refrigerant piping of suitable diameter with proper insulation run from out door unit to individual indoor unit Supported proper slotted trays and Y connections are used to connect piping with outdoor & indoor unit.

1.18.5 Duct and Air Outlet: The indoor unit is provided with proper air outlet and ducts are kept far from the room to be air conditioned & if required to blow the air. The provisions for shape, size and material are made as in case of Chiller plant.

1.18.6 Wiring: The cable and control TPN switch of proper size are used to feed power supply to the out door units and proper submain is used to feed power supply to indoor unit.

1.19 Advantage of VRF / VRV System over Chiller Type Air Conditioner:

1. The VRF system is very compact system and do not require more space for long water pipe and ducts etc as in case of chiller type plant.

2. It is instant start, the whole plant can be started by a remote by any individual indoor unit or out door unit, while chiller plant is got to start at feast one hour earlier to achieve proper temperature and humidity. Thus it save energy with not using in advance to start.

3. It is variable refrigerant flow volume (VRF or VRV) i.e. the volume of refrigerant required is reduced as per required reduced heat load, i.e. if some rooms or portion of buildings is to be air conditioned, the volume of refrigerant required is less, unlike in chiller system in which whole refrigerant is used irrespective of load required thus 3PH electrical energy is saved.

4. The VRV system uses Ozone friendly refrigerant while chiller plant uses internationally banned ozone effective R-22 refrigerant.

5. No operator is required to run VRV system as incase of chiller system.

6. It has least maintenance problem as compared to chiller plant and it enables quicker problem resolution and less down type.

7. Outdoor unit can be installed any where even on top of the building to reduce pipe length while in chiller plant, the compressor, condenser & evaporator are required to be installed at distances from the building.

8. VRV system is noiseless as compared to chiller system which has great noise due to vibrations using big motor.

9. In VRV system, heat is transferred directly from air to refrigerant, while in chiller system water is used as medium to transfer heat from air to refrigerant.

10. VRV system can be used on reverse air conditoning cycle for heating, while hot water, steam or heat elements are used in place of chilled water to heat air. Further VRV system requires less time in installation and as it can assemble easily. The scroll compressor has

further more advantage in comparison with inverter type compressor i.e. power energy saving, high efficiency and system automatically switches over between heating and cooling. Scroll compressor has no magnetic interference.

11. VRV system has high comfort as each indoor unit can be controlled individually to meet occupant comfort. Further it has precise temperature control.
12. VRV has higher reliability and wide variety of indoor units.
13. It is free from oil recycling equipment system so long & large height of pipe can be used with no effect on the air conditioning
14. It has no resonance or frequency noise so large high air volume fan can be used.
15. Scroll compressor electric expansion value gives better refrigeration effect. Scroll compressor has high heat transfer effect using expansion 3 row inner grooved heat exchanger.

1.20 Comfort Condition :

We design heat load for following comfort conditions.

Climate	Effective Temperature °C	Corresponding DB At 50 % RH
Hot and dry (April & May)	21.1 to 26.7	23.9
Hot and humid (Mansoon)	22 to 25.6	26.9

In general the optimum value of comfort condition are taken as

ET   21°C
DBT   25± 1°C
RH   50 ± 5 %

1.21 Design of Capacity of Air Conditioner:

1.21.1 Heat Load: To calculate the size or capacity of air conditioner or air conditioning system or the refrigeration system keeping in view of prevailing inside & out side condition, we must know the exact amount of heat to be removed or added to maintain required inside condition i.e. dry bulb temperature & humidity. Total amount of heat required to be removed or added from or to the space to maintain desired temperature in the space by the air conditioner is called heating load or cooling load of air conditioning or refrigeration equipment either in summer or winter i.e. when inside temperature of space or room is greater or less than the desired temperature.

1.23.1.1 Heating Load or Heat Load During Summer: Let us consider the case that inside temperature is greater than desired temperature in summer , so we discus and find out the source to give heat or increase inside temperature to calculate the heat load i.e. total heat to be removed. The case to find out heat to be added in winter to find cooling load is just reverse to that in summer assumed above.

There are two main component of heat load in summer.

(i) **Sensible heat gain:-** When there is a direct addition of the heat to the enclosed space or room, a gain in sensible heat occur. This sensible heat gain may occur due to any or all the following source of heat transfer.

(a) The heat flowing into the building by conduction through walls, floors, ceiling, doors and windows due to temperature difference on their two sides.

(b) The heat received from solar radiation. It consists of.

(i) Heat transfer or transmitted heat directly through glass of window ventilators or doors.

(ii) The heat absorbed by walls and roof exposed to solar radiation and later on transferred to the room by conduction.

(c) The heat conducted through interior partitions from rooms in the building which are not to be air conditioned.

(d) The heat given off by light, motors, machinery, cooking operation industrial processes etc.

(e) The heat liberated by the occupants.

(f) The heat carried by the outside air leakage in (infiltrating air) through cracks in doors, windows or through their frequent opening.

(g) The heat gain through walls of duct carrying conditioned air through the conditioned space in the building.

(h) The heat gain from the fan work.

(ii) **Latent heat gain:** Where there is an addition of water vapor in the air of enclosed space, a gain in latent heat may occur due to any or all the following sources.

(a) The heat gain due to moisture in out side air entering due to infiltration.

(b) The heat gain due to condensation of moisture from occupants.

(c) The heat gain to moisture passing directly into conditioned space through permeable walls or partition from outside or from adjoining regions, where vapor pressure is higher.

The total heat load to be removed by air conditioning or refrigeration equipment is the sum of sensible heat and latent heat load as discussed above.

The total heat load is converted in to ton of refrigeration unit to find the capacity of air conditioning equipment.

1.21.2 Heat Load Calculation: Heat load by various factor can be calculated by following steps.

(1) **Product load:** The product is at high temperature and its temperature is to be reduced for its preservation for long time e.g. medicine, food, drinks. The heat gain depends on the weight of product (m), temperature difference of product with final temperature up to which product to be cooled.

Heat gain = MS (T2 - T1) = MS ?T

(2)Air circulation load: Opening of door and window shutter of the air conditioned or storage room / cabin, air at high temperature enters inside. Knowing infiltration of air in building heat load can be worked out.

Air change / hr are taken for office and auditorium as 1 or 5 respectively.

$$\text{Amount of infiltrated air} = L \times W \times H / 60 \, Ac \; ^{m^3}/_{min}$$
$$= V \times Ac / 60 \text{ where}$$
$$Ac = \text{Air charge per hour}$$

L, W, H & V are length, width, height & volume of room

(3) Miscellaneous load:

3(a) Solar heat load: Sun is bigger source of heat transmission by conduction convection or radiation. Heat is also received by air in side room by radiation through glass or ventilator or by conduction through wall and ceiling of building.

Heat flow by convection & radiation through glasses	= Absorbed solar radiation + Radiation and convection heat exchanged between glass and outer surface

This can be found from the guide published by (ASHVE) American Society of Heating and Ventilating Engineers. 35% heat gain through glass can be reduced by using curtains on glass, while 20% heat can be further reduced by using heat absorbing or vanishian or painted glass, while double pane glass can reduce further 10% heat load. If Shades are made on window or door out side building 7.5% solar heat load is reduced.

3(b) Wall / ceiling heat load: Solar heat via conduction through walls is called walls & ceiling is called walls / ceiling heat load. This depends on the type and thickness of insulation, surface area of walls or ceiling and temperature difference between outside and inside temperature.

$X_1 \quad X_2 \quad X_3$

Heat load Q = UA ΔTe

Where U = conduction coefficient

A = surface area

ΔTe = temperature difference

U = coefficient of heat transmission through wall

= $1 / \{1/f_o + x/k + 1/f_i\}$

If wall is made of different material, then over all coefficient of heat transmission or conductance coefficient is given by.

$$U = \frac{1}{1/f_o + x_1/k_1 + x_2/k_2 + \underline{\quad} + 1/f_i}$$

Coefficient of conductance and equivalent temperature difference can be found from the table given in various book on air conditioning or given by manufacturer of air conditioner for glass, walls, ceiling etc. separately.

3(c) Electrical energy dissipation heat load: The place to be air conditioner may have heat dissipation source of electrical energy e.g. lamp, motor or electrical appliance etc. heat gain through these source can be worked by following data of dissipation energy.

Electrical lamp	= 1.9 K cal / watt
Fluorescence tube lamp	= 1.89 K cal / watt
Electric motor	= 1.610 K cal / HP
Other electrical appliance	= 0.0056 K cal

3 (d) Heat gain through duct: Heat gain due to supply duct if inside the space to be air conditioned depends upon the temperature of air in duct & temperature surrounding duct.

Heat gain	$= U A_D \, \Delta t e$ when
U	= over all heat transfer coefficient
A_D	= surface area of duct

(4) Occupancy or people load: The human body in cooled space constitute both sensible and latent heat load. These heat gain are different at different place depending on density of occupancy. The values of sensible & latent heat load per person in different building is given in a table supplied in air conditioning book or by manufacturer of air conditioner.

Name of building	Sensible heat	Latent heat
Theater	108 Kcal / hr	80 Kcal / hr
Office	111 Kcal / hr	113 Kcal / hr
Dance hall	135 Kcal / hr	130 Kcal / hr

1.21.3 Heat Load Estimation During Winter: In the similar manner the cooling load or heat load estimation during under is made, on account of maximum probable heat loss of room or space to be heated, so the heating system plant in designed that it has a capacity to meet the heat load required to supply heat loss inside room or space due to following factor with respect to comfort condition.

(1) Transmission heat loss: Heat loss due to four walls & roof of the room is calculated on

the basis of DBT temperature.

$$Q = UA (t_1 - t_2)$$
$$= UA\Delta T$$

(2) Solar radiation heat loss: There is no solar radiation present.

(3) Internal heat gain: Internal heat gain from occupants light, motor and machinery, appliance etc, diminish heat requirement. So this negative heating load is accounted only after careful consideration. The size or use of the space/room to be air conditioned is important factor to keep is mind at night, week ends or other unoccupied period.

1.22 Design of Heating Plant:

So the heating plant capacity is designed sufficient to bring inside temperature to the desired value i.e. comfort condition, before occupant comes.

Heating the air inside room or space is also done by same summer air conditioning plant working on reverse cycle. So the air conditioner plant working with one refrigerant, for cooling during summer, air conditioning work on reverse cycle for heating, so the hot air supplied inside the room white the air loosing heat (cold air) is exhausted out side.

So we design for heating load & select the capacity of air conditioning plant for summer air conditioning and install the plant, which is also used for heating.

1.23 Example:

The data of top multi-storied building is below

(i) Area of glass of door & widows on all four side North, East, South, West are 60 m², 32m², 60 m², 32 m², total glass area   =184 m².

(ii) Area of walls on all four side North 102 m², South 102 m², East 72 m², West 72 m².

(iii) Area of partition wall or wall in side shade   = 230 m²

(iv) Area of walls & ceiling exposed to sun   = 1103 m²

(v) Fresh air infiltration   = 70 m³/mm

(vi) Occupancy   = 250 persons

(vii) Light and electrical appliance   = 11825 watt

Item	Area of Quantity	Temperature difference of equivalent temperature difference	Factor	Heat gain BTU / hour
Room sensible heat				
Solar gain glass				
N glass	60m²	91	—	5460
S glass	60m²	32	—	1920
E glass	32m²	32	—	1024
W glass	32m²	492	—	15774
			Total	24178
Solar transmission gain wall & cell				
N wall	102 m²	12.2	3.5	4355.4
S wall	102 m²	18.3	3.5	6533.1
E wall	72 m²	18.3	3.5	4611.6
W wall	72 m²	17.4	3.5	4384.8
Roof exp	(1103-184) 919m²	30.6	2.12	64807.1
Roof exp shade	—	—	—	—
			Total	84692.0
Transmission gain except wall & celling				
All glass	184 m²	18	59	19540.8
Partition	230 m²	155	1.86	8660.9
			Total -	28201.7
Internal gain				
Walls/ceiling in shade	919 m²	2.5	6.5	14933.8
Infiltration	70 m³/mm	18	204	25204
			Total	40137.8
Internal heat				
People	250	-	75	18,750
Motor	-	-	-	-
Light	11825 watt	-	1.25	18451.2
			Total	29701.2
			Total room sensible heat -	206910.7
			Heat gain due leakage loss -5%	10345.5
			Effective room sensible heat -	217256.2
Room latent heat				
Infiltration	70 m³	006	50,000	21,000
People	250	-	35	8750
Appliance	-	-	-	-
			Total	29750

Find out total heat load and design from proper air conditioner plant.

Solution   Calculation chart

Total room latent heat   = 29750

Leakage load factor 0. 5%   = 437.5

Effective room latent heat   = 30,182.5

Total room heat load / gain   = 2,47,438.7

Duct side air heat gain:

Sensible heat   70 m³/min   204 (1- 0.15)   = 21,848.4

Latent heat   70 m³/min   50,000 (1- 0.15)   = 17850.0

              Total duct side   = 39698.4

              Total heat load   = 287137.1W

                            = 285KW

              Other loss pipe etc. 3%   = 8.61KW

              Grand total of heat load   = 295.6KW

                            = 295.6/3.5

                            = 84.5TR

Sensible heat factor and apparatus dew point (ADP)

ESHF = Room sensible heat / Room total heat

        = 217256.2/ 247438.7 = 0.88

ADP = Indicated 25°C Selected   For psychrometric chart TADP=25°C

              Dehumidified Air Quantity

Temperature rise = (1 - BF) (TRM - TADP) = 0.85 (25- 12) = 11.05°C

Dehumidified Air = Room sensible heat / 1.08 x temperaturerige

              = 217256.2 / 1.08 x 11.05

              =18205m³/min

from above calculation it is evident that air conditioner of 84.5 TR ton capacity is required and 18205 m³/m dehumidified air is required for above plant. Accordingly to this plant capacity and dehumidified air, the specification of following component of the air conditioning chiller plant can be chosen.

(A) Compressor (B) Condenser & its pump (C) Evaporator (chiller) and its cooling tower (D) Air handling unit etc. together with pipe and duct.

Example 2: An air conditioner system is designed for a restaurant where the following data is available.

Total heat flow through the walls   = 6.2 KW

Roof and floor   = 2 KW

Equipment sensible heat gain   = 2.9 KW

Equipment latent heat gain	= 0.7 KW
Total infiltration air	= 400 m³ / hr
Out door condition	= 31°C DBT 26°C BT
Inside designed condition	= 27°C DBT 55% RH
Minimum temperature of air supplied to room	= 17°C DBT
Total amount of fresh air supplied	= 1600 m³ / hr
Seating chairs for dining	= 50
Employees serving the meals	= 5
Sensible heat gain per person	= 58 W
Latent heat gain per seating person	= 44W
Latent heat gain per employee	= 76 W
Sensible heat added from meals	= 0.17 KW
Latent heat added from meals	= 0.3 KW
Motor power connected to fan	= 7.6 KW

If the fan is situated before the conditioner then find the following

(a) Amount of air delivered to the room m³ / hr.

(b) Percentage of recirculated air.

(c) Refrigeration load on the coil in tones of refrigeration.

(d) Dew point temperature of the cooling coil by pass factor.

Solution : Given G_{SW} = 6.2 KW, Q_{SG} = 2 kw, Q_{SE} = 0.7 kw, Vi = 1600 m³ / h, td_1 = 35°C Tw_1 = 26°C, td_2 = 27°C, Φ_2 = 55%, td_4 = 17°C, Vr = 1600 m³ /h, dining chair = 50, employees = 58, Q_2 per person = 58 W, Q1 per person = 44W,

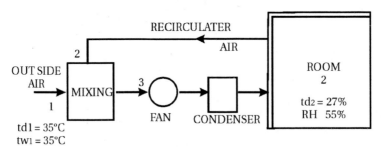

Block Diagram

Q_1 per employees = 76 W, Q_{SM} = 0.17 KW, Q_{CM} = 0.3 kh, Q_M = 7.6 KW

out side condition of air i.e. 35°C DBT and 26°C WBT (point - I)

Inside design condition of air 27°C DBT and 55% RH (point - II)

From psychometric chart of specifies volume of air at point - I

VSI	= 0.897 m/Kg of air
Enthalpy at point -I h_1	= 80.6 KJ of dry air
Enthalpy at point - 2 h_2	= 2KJ / Kg of dry air
Enthalpy at point A h_A	= 66 KJ / Kg of dry air

Mass og infiltration $M1 = V_1 / V_{SI} = 400 / 0.897 = 446$ kg/hr

Sensible heat gain due to infiltration $= Vh_1 (h_4 - h_2)$

$$= 446(66-58.2) = 3480 \text{ KJ/hr}$$
$$= 3480/3600 = 0.97 \text{ Kw}$$

And latent heat gain due to infiltration $= m_1 (h_1 - h_4)$

$$= 446 (806-66)$$
$$= 6512 \text{ KJ/hr}$$
$$= 6512 /3600$$
$$= 1.8 \text{ KW}$$

Sensible heat gain from person $= Q_s$ per person x No. of person

$$= 58 \times 50$$
$$= 2900W$$
$$= 2.9 \text{ KW}$$

Latent heat gain person $= Q_1$ per person x No. of person

$$= 44 \times 50$$
$$= 2200W$$
$$= 2.2 \text{ KW}$$

Sensible heat gain from serving employees $= Q_s$ per person x No. of person

$$= 58 \times 5$$
$$= 290 \text{ W}$$
$$= 0.29KW$$

Latent heat gain from serving employee $= Q_1$ per person x No. of person

$$= 76 \times 5$$
$$= 380 \text{ W}$$
$$= 0.38 \text{ KW}$$

Total sensible heat in room = Heat gain from wall,roof and floor Q_{SE}

+ Solar heat gain through glass Q_{SE}

+ Sensible heat gain from equipment Q_{SE}

+ Sensible heat gain due to infiltration air

+ Sensible heat gain from person taking meal

+ Sensible heat gain from employee serving meal

$$= 6.2+2K_w+2.7KW+0.97+2.9+0.29+0.17$$
$$= 15.43 \text{ KW}$$

Total latent heat in room	= Latent heat from equipment
	+ Latent heat from infiltration air
	+ Latent heat from person taking meal
	+ Latent heat from serving employee
	+ Latent heat from meal
	= 0.7+1.8+2.2+0.38+0.3
	= 5.38 KW
Room sensible heat factor RSHF	= RSH / RSH +RLH
	= 15.43 / 15.43 +5.38
	= 0.741

Mark RSH = 0.741 on sensible heat factor scale as point a and join with point b 26°C DBT, 50% RH on alignment scale from point 2 draw a live 2-5 parallel to line ab. The 2-5 line is known as RSHF line. Now a vertical line through 17°C DBT (minimum temperature of air supplied to room) which cuts RSHF line at point 4 from the psychometric chart we find specific volume of air supplied at point 4.

V_{S4}	= 0.36 m/Kg of dry air
Enthalpy of air point 4 h4	= 45 KJ / kg of dry air
(a) Amount of air delivered to room ma	= Total room heat / total heat removed
	= (RSH+RLH) / (h_2 - h_4)
	= (15.43 + 5.38) / (58.2 - 45)
	= 1.576 kg /s
	= 1.576 x 3600
	= 5675 kg / hr
Volume of air deliver to room V_d	= ma x VS4
	= 5675 x 0.836
	= 4745 m/h

(b)Percentage of recirculated air mass of air

Supplied mF	= V_F / V_{SI}
	= 1600 / 0.817
	= 1784 kg/hr
Mass of recirculated air	= ma - SI
	= 5675 - 1784
	= 3890 kg/hr
Percentage of recirculated air	= 3891 x 1 / 5675 x 100
	=68.6%

(c)Refrigerator load on the coil

Recirculated air st point 2 is 68.6% so fresh air supplied is 31.4% air (fresh & recirculated) enter the cooling coil say point 3 on psychometric chart, such that length 2-3 = length 1-2 x 0.314

Enthalpy of air at point 3 h3   = 04.6 KJ/1kg of dry air

Refrigeration load on the coil $= m_a(h_3 - h_4)$ + heat added

         = 5675 (64.6 -45)+7.6

         = 38.5 KW

         = 38.5 / 3.5

         = 11 TR

(d) Dew point temperature of cooling coil & by pass factor

Joint point 3 with point 4 and produce the line to intersect the saturation curb at point 6 from psychometric chart.

Tdp $= td_6 - 14.6°CA$

(e) Now by pass factor BPF $= (td_4 - td_6) / (td_1 - td_6)$

         = 17-14.6 / 29.5-14.6

         = 0.161

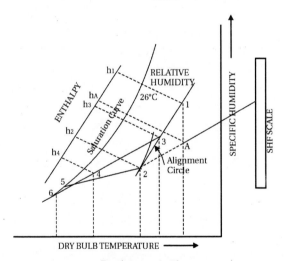

Psychrometric Chart

1.22 Estimating & Costing:

1.24.1 Design of air conditioning unit:

1.24.1.1 Psychromatic chart method: We can find out heating load or cooling load by knowing various parameters discussed earlier for any room space or complete building for selection or unitary or central air conditioning system.

The various parameter to be know as.

(i) Dimensions of room / space.

(ii) Wattage of electrical appliances like luminaries, electric fans, pumps, T.V, refrigerator etc.

(iii) Maximum number of person likely to be present (maximum occupancy).

(iv) Fresh air

(v) Outside and inside temperature of room/premises.

(vi) Psychometric chart & relevent table.

1.24.1.2 Software Method: Software are available by which we can find out heating load /cooling load, by feeding required known parameter in the computer loaded with requisite software Knowing heating/ cooling load required of required air conditioning unit i.e. TR, the capacity of various unit of air conditioner i.e. Condenser,Compressor, Evaporator. Cooling tower, AHU, duct, air circulating fan etc are selected

1.24.1.3 Thumb Rule Method: Unitary air conditioners can also be designed by thumb rule. Selection of unitary unit only for individual room can be found by thumb rule, knowing the volume of space and average occupancy.

Volume of room	Occupancy	Ton capacity of one unit
40 m³	6 to 8	1.5 TR
50 m³	10 to 12	3.0 TR or two 1.5 TR
60 m³	15 to 20	4.0 TR or two 2.0 TR

1.24.1.4 Detail Of Measurement: After design of TR capacity of air conditioner & quantity of dehumidify air, the number and size of air conditioners are selected and if central air conditioning is required, then after designing the plant capacity & dehumidify air, chiller plant, and its compressor, condenser, evaporator, AHU & duct fan etc are selected. After selection of size of the component units, their number with complete specification is summarized in a table called detail of measurement..

1.24.1.5 Analysis of Rate: Proper analysis of rates of each item is prepared as per standard by obtaining the rates of each items from manufacturer of plant as per price list or current market rates.

1.24.1.6 Bill of Quantity: Knowing the quantities and rates of various items required to be executed, a bill of quantity on standard format is prepared having items with brief specification, their quantities, rates and amount. The total of amount column will give total coast of Supply, Installation, Testing and commissioning (SITC) of air condition system work required for execution

1.24.1.7 Abstract of Cost: Abstract of cost is also prepared as per standard norms including the expenditure of miscellaneous items, contingencies together with the cost of lump-sum provision of civil construction etc, if required and centage charges, if required. The total of the abstract of the cost will give the cost of the estimate or DPR etc.

2 | Lightning and Its Protection

2.0 Introduction

Prior to eighteen centaury people used to consider the lightning as God's way of venting or expressing their anger on mankind and men were always been terrified and fascinated by it. Benjamin Franclin in 1753 through his kite experiment demonstrated the existing lightning discharge resulted due to lightning. Dalibord at almost same time confirmed the theory with his metal rod experiment, which was electrified during lightning storms.

Later on the scientists put forward theory of charged cloud striking each other which produces charged flash or bright light i.e. lightning together with huge amount of heat and sound, during thunder storm. The lightning, when comes in contact with any object it discharges through it.

It could not yet possible to stop this natural phenomenon. Thunderstorm, thunder-stroke is resulting lightning, large noise and huge heat. Production of heat energy increases the temperature of the atmosphere resulting in a strong air pressure wave and damage, the object which comes in contact with it. Production of great excessive noise (Thunder) may affect sound pollution and other adverse effect on human ear and huge light is produced during lightning (optical effect).

However lightning produces much serious and adverse effect and causes damages due to electrical discharge in comparison with heat and sound energy. There is no protection against heat or sound. The lightning discharge damages to the object through which it discharge. Therefore it is essential to provide protection against lightning, discharge, which is possible.

If the object is a person, whose occupation keep him outdoor, the lightning hazard is greatest but the proportion of injury in general is very small, if the person is inside a building with proper protection scheme. The practice, design, installation, testing and maintenance of protective scheme for the building are covered vide IS code no. 2309.

Though a lightning protection system duly designed and installed in accordance with all norms, can not guarantee absolute protection to structures, person or objects, however it will significantly reduce the risk of protected structure being damaged by lightning.

2.1 Definition:

2.1.1 Lightning: During the natural phenomenon thunderstorms or thunder stroke, the creation of flash or bright light together with huge sound and heat is called lightning.

2.1.2 Lightning Flash to Earth: An electrical discharge of atmospheric origin between cloud and earth consisting of one or more current impulses (return strokes) is called lightning flash to earth.

2.1.3 Lightning Stroke: One or more lightning discharges to earth is called lightning stroke.

2.1.4 Striking Point: A point where lightning stroke contacts the earth, a structure or lightning protection system is called striking point.

2.1.5 Lightning Flash Density Ng: Yearly number of lightning flashes per km² is called lightning flash density.

2.1.6 Return Stroke Density Na: Yearly number of return stroke per km² is called return stroke density. A lightning stroke consists in average of several return strokes.

2.1.7 Lightning Protection System (LPS): The complete system used to protect structure and open areas against the effects of lightning is called lightning protection system. It consists of (ELPI), External lightning protection installation and an internal lightning protection installation (ILPI) if any.

2.1.8 External Lightning Protection Installation (ELPI): An external lightning protection installation consists of an air termination system, down conductor and earth termination system.

2.1.9 Internal Lightning Protection Installation (ILPI): An internal lightning protection installation consists of all the devices and measures reducing electromagnetic effect of light current with in protected volume.

2.1.10 Protection Volume: It is defined as volume of influence of the early streamer emission lightning conductor with in which ESE lightning conductor is the striking point i.e. light discharge touch ESE terminal.

2.1.11 Protection Zone: Zone of protection of a lightning conductor denotes space with in which protection against a direct lightning stroke is provided.

2.1.12 Lightning Conductor: It is a conducting path of least resistance between general mass of earth and atmosphere above the structure to allow the lightning discharge to enter earth.

2.1.13 Air Termination or Terminal: The outermost part of lightning conductor which is striking point of lightning discharge is called air terminal or air termination

2.1.14 ESE Terminal: It is a lightning rod terminal equipped with a system which creates triggering advance of the upward leader when compared with simple rod (SR) lightning conductor in the same conditions.

2.1.15 Triggering Process: Physical phenomenon between the inception of the first corona and the continuous propagation of an upward leader is called triggering process.

2.1.16 Triggering Advance: It is mean gain in triggering time of the upward leader of the ESE lightning conductor, when compared with a S.R lightning conductor in the same condition and derived from the evaluation test. This is expressed in μ s.

2.1.17 Natural Component: It is conductive part located outside the structure, sunk in the walls or situated inside a structure and which may be used to replace all or part of a down conductor or as a supplement to an ELPI.

2.1.18 Equipotential Bonding Bar: It is a collector used to connect the natural components, ground conductor, earth conductor, screens, shields and conductor protecting electrical telecommunication lines or other cables to the lightning protection system.

2.1.19 Equipotential Bonding: An electrical connection putting ground conductors and conducting parts at the same potential or a substantially equal potential is called equipotential bonding.

2.1.20 Equipotential Conductor: A conductor providing for equipotential bonding is called equipotential conductor.

2.1.21 Dangerous Sparking: An electric are produced by a lightning current within the volume to be protected is called dangerous sparking.

2.1.22 Safety Distance (S): The minimum distance for which no dangerous spark can be produced is called safety distance.

2.1.23 Interconnected Reinforcing Steel: It is a natural components within a structure which provide an electrical path of resistance smaller than $0.01\ \Omega$ and can be used as down conductors.

2.1.24 Down Conductor: Part of the external lightning protection installation designed to conduct current from lightning conductor to the earth termination system is called down conductor.

2.1.25 Test Joint/Disconnect Terminal or (Measurment Terminal): It is a device used to disconnect the earth terminal system from the remainder of the lightning protection system.

2.1.26 Earth Electrode: A part or group of part of the earth termination system which provides direct electrical contact with the earth and disperses the lightning current to earth is called earth electrode.

2.1.27 Earth Termination System: A conductive part or a group of conductive parts in intimate contact with earth which provides an electrical connection with earth is called earth terminal system.

2.1.28 Earth Termination System Resistance: Resistance between the test joint and earth is called earth termination system resistance. It equals the quotient of potential increase, measured at the test joint with respect to an infinitely remote reference and of the applied current to the earth electrode.

2.1.29 Surge Protective Device (S.P.D): A device designed to limit transient surge voltage and to provide a path for the current waves is called surge protective device. It contains at least one non linear component

2.1.30 Transient Surge Voltage of Atmosphere Origin: It is over voltage lasting a few milliseconds only, oscillatory or not, usually strongly damped.

2.1.31 Protection Level: Classification of a lightning protection system which expresses its efficiency is called protection level.

2.1.32 Equivalent Collection Area of a Structure: A flat ground surface subjected to the same number of lightning flashes as the structure under consideration is called equivalent collection area.

2.2 Characteristic of Lightning Discharge:

Lightning discharge to a structure are unidirectional current discharge through structure, which may very from few hundred ampere to 200 KA in magnitude as shown in figure 2.2 and rises in microsecond to crest value and falls to zero in milliseconds. It may be single or multi stroke which follow the same path. The discharge is of the form of surge or pulse of current and last for a second or few seconds.

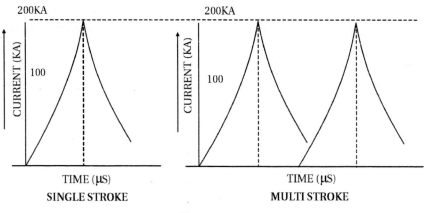

Fig 2.2

Lightning is mainly characterized by parameter related to the electric arc between the cloud and the ground i.e. related the lightning current (discharge) or as discussed earlier. The important lightning characteristic parameter are Amplitude, Rise time, Decay time, Current variation rate di/dt, Polarity, Charge, Specific energy, Number of strike / stroke per discharge. The parameter amplitude, rise time and decay time are independent of statistic i.e. discharge of any amplitude, with any rise or decay time may be encounted.

2.3 Principal Effect:

There are three principal effect of lightning discharge on building beside optical acoustical effect.

(i) **Electrical effect:** The lightning current, when discharge through structure, produce high voltage gradient due to high impedance of structure which may be dangerous to the living being in building. Prior to current stroke, electromagnetic radiation produces flash over and induction. Because of this huge discharge current chemical change of material (electro-chemical effect) with fusion or melting may occur.

(ii) **Thermal effect:** A high lightning current passes through structure for a short time resulting rise in temperature due to heating, it may or may not give fusing of iron structure or reinforcement of building, but it ruptures the structure certainly. Excessive rupturing may result in fusing by further heating.

(iii) **Electrodynamical or mechanical effect:** When a high current is discharged through parallel conducting path of structure situated in closed proximity, it is subject to large mechanical force which may damage the structure.

2.4 Protection and Protection System:

The fundamental principle for protection against lightning is to provide conducting path of least resistance between general mass of earth and atmosphere above the structure to allow the discharge to enter earth, without passing and damaging structure made of brick, stone or concrete. This path is called lightning conductor.

In 1753 Benjamin Franklin invented lightning conductor. In 1880 Belgian physicist Miellsen recommends the protection of building by covering them with metal wire connected to a series of spike on the roof and then well earthed. This meshed cage was first used as protection device or lightning conductor. Then later on finial rod or strips on edge or projection of building were used as lightning conductor. Down conductor and earthing was necessary. Since then till now they are used in the conventional system of lightning conductor. A high potential voltage surge is developed at the point of strike if the lightning spark touches the air terminal. If the lightning do not touch air terminal a high voltage is induced in the air terminal. The high lightning surge voltage or induced voltage discharges or passes to earth through down conductor. In 1986 after research of several years, a new type of lightning conductor was developed giving strong ionization around the terminal and is called ESE (Early Streamer Emission). The high voltage resulted due to strong ionization results in emission of streamer and discharges through down conductor to earth like in old conventional air terminal. The lightning conductor may be adequately earthed metal part of least resistance which are properly spread over building. The thermal effect may be reduced by providing conductor of least cross section. The long metal water pipe, metal sheath or armoring of electrical cable etc. buried but not connected to lightning conductor, remain on earth potential and since the point of strike of lightning conductor may be raised to high potential, there is risk of flam over from high potential to low potential. To safeguard this lightning conductor is placed on upper most of building or its projection. Equipotential bonding of metal part of building is done. The range of lightning conductor over which lightning discharge is attracted is not constant but is statistical and proportional to serenity of discharge and is negligibly affected by configuration of lightning conductor i.e. vertical or horizontal air terminal are equivalent, so unless for practical consideration any type of air termination is used.

2.5 Material and Their Dimension:

Copper has good conductivity and least resistance, so it is preferred as material for vertical finial rod or horizontal strip as air terminal. The material of lightning conductor i.e. air terminal, down conductor & earth terminal must be sufficiently strong, good con.' ¬tor and adequately protected against corrosion. Following material and its shape is recommended.

Material & the shape	Conductivity	Strength	Prodection against corrosion	Cost
1. Solid or stranded copper wire or strip	90%	Good	Bad	More
2. Copper covering or effective welded steal cage	70%	Better	Sufficient	Least
3. Galvanized or Zink coated steal	Further low	Best	Good	Least
4. Aluminum wire or strip	Double that of copper of same weight	Sufficient	Effective	Lesser
5. Aloy	Fairly good	Fair	Substantial	Moderate

Recommended shape and minimum size of conductor for use on above ground or below ground are give as below.

Above ground			Below ground		
Sl. no	Material & shape	Minimum size	Sl. No	Material & shape	Minimum size
1.	Round copper or copper clad steel wire	6mm dia	1.	Round copper wire or copper clad steel wire	8mm dia
2.	Stranded copper wire	50mm² dia (7 x 3.0 mm dia)	2.	------------	-----------
3.	Copper strip	20 x 3mm	3.	Copper strip	32 x 6mm
4.	Round GI wire	8mm dia	4.	Round GI wire	10mm dia
5.	GI strip	20 x 3mm	5.	GI strip	32 x 6mm
6.	Round Aluminum wire	5mm dia	6.	-----------	------------
7.	Aluminum strip	25 x 3.0mm	7.	-----------	------------

2.6 NEE of Protection

2.6.1 Risk Index Method: Based on experience, a digital system for assessing the risk of damage to building due to lightning is uniformly adopted. An index figure is allotted to the various factors influencing the risk of damage to building in this system and we sum up the index figure due to all various factors to determine whether or not the protection is needed at all. The total figure thus obtained by summing up all index figures are called RISK INDEX. Higher the risk index, greater is the need for lightning protection. For practical utility of this valuation safe risk index may be taken as forty (40) based on experiences. However this system is not regarded as sole criterion but taken as an aid to judgment.

For example the structure, made entirely of metal with adequately earthed, are self protecting, while the building with frame of steel or reinforced concrete can easily be protected with less cost. On the other hand the building with larger people density or used for essential public service or situated in area prevalent lightning stroke or very tall or isolated or using highly inflammable explosive material or of grater historically or cultural importance need lightning protection necessarily.

Various factors and their risk index figures are given in following table.

Table No. 1 Index figure for uses of structure.

Sl. No.	Use of structure	Value of index figure
1.	Houses or buildings of comparable size.	2
2.	House or buildings of comparable size with outside aerial.	4
3.	Small or medium size factories, works and laboratories.	6
4.	Big industrial plants, telephone exchanges, office blocks, hotels, blocks of flats and residential buildings.	7
5.	Places of assembly e.g. churches halls, theatres, museums, exhibitions, department stores, stations, airports and stadiums structures.	8
6.	Schools, hospitals, children's home.	10

Table No.2 Index figure for type of construction.

Sl. No.	Type of construction	Value of Index figure
1.	Steel framed encased with any non-metallic roof.	1
2.	Reinforced concrete with any non-metallic roof.	2
3.	Brick, plain concrete or masonry with any roof other than metal or thatch.	4
4.	Steel framed encased or reinforced concrete with metal roof.	7
5.	Timber framed or clad with any roof other rain metal or thatch.	7
6.	Brick, plain, concrete or masonry, timber frame but with metal roofing.	8
7.	Any building with thatched roof.	10

Table No. 3 Index figure for contents or consequential effect.

Sl. No.	Contents or type of building	Value of Index figure
1.	Ordinary domestic or office buildings, factories and workshops not containing valuable or especially susceptible contents.	2
2.	Industrial and agricultural buildings with valuable or especially susceptible contents.	5
3.	Power stations, gas works, telephone exchanges radio stations.	6
4.	Industrial key plants, ancient monuments and historical buildings, museums, art galleries or other buildings with especially valuable contents.	8
5.	Schools, hospitals, children's homes and other such homes, places of assembly.	10

Table No. 4 Index for figure degree of isolation

Sl. No.	Degree of isolation	Value of index figure
1.	Structure located in a large area of structures or trees of same or greater height (e.g. large town or forest).	2
2.	Structure located in an area with few other structure or trees of similar height.	5
3.	Structure completely isolated or exceeding at least twice the height of surrounding structure or trees.	10

Table No. 5 Index figure for type of terrain.

Sl No.	Type of terrain	Value of index figure
1.	Flat terrain at any level.	2
2.	Hill terrain.	6
3.	Mountain terrain between 500m to 1000m.	8
4.	Mountain terrain above 1000m.	10

Table No.6 Index figure for height of structure.

Sl. No.	Height of structure above ground		Value of index figure
	Exceeding (m)	Not Exceeding (m)	
1.	-	10	2
2.	10	15	4
3.	15	20	5
4.	20	25	8
5.	25	30	11
6.	30	35	16
7.	35	40	19
8.	40	45	22
9.	45	55	30
Structure exceeding 55m require protection in all case.			

Table No. 7 index figure for lightning per year.

Sl. No.	Number of thunder storms days per year		Value of index figure
	Exceeding	Not Exceeding	
1.	-	5	4
2.	5	10	8
3.	11	15	13
4.	16	20	18
5.	21	-	21

Table No. 8 Index figure for place showing number of thunder storm days per year.

Sl. No.	Name of place	Annual thunder storm days	Sl. No.	Name of place	Annual thunder storm days	Sl. No	Name of place	Annual thunder storm days
1.	Mussuorie	61	8.	Agra	24	15.	Jhansi	20
2.	Roorkee	74	9.	Mainpuri	23	16.	Allhabad	51
3.	Najibabad	36	10.	Bahrich	31	17.	Varanasi	51
4.	Mukteshwar	53	11	Gonda	22	18.	Azamgarh	1
5.	Meerut	-	12.	Luknow	18	19.	Gorakhpur	11
6.	Barreilly	34	13.	Kanpur	26			
7.	Aligarh	30	14.	Fatehpur	24			

Examples showing calculation of risk index factors for various types of buildings and need of protection is illustrated in the following table.

Sl. No	Example	Usage of structure	2 Type of construction	3 Contents	4 Degree of isolation	5 Type of terrain	6 Height of structure	7 Lightning prevalance	Risk index total figure sum of 1 to 7	Recommondation
1.	Small residential building in a thickly populated locality (height less than 10m)	2	4	2	2	2	2	21	35	Not recommon-dat
2.	Office building in thickly populated locality (height 20m)	7	2	2	2	2	5	21	41	May or not use protection depending upon the building
3.	Hotel building (height 31m) exceeding height of surrounding structure	7	2	2	10	2	16	21	60	Protection essential
4.	Building of historical importance completely isolated (height 50m)	8	4	8	10	2	30	21	83	Protection essential
5.	Structure of high historical importance (height exceeding 55m)	-	-	-	-	-	-	-	-	Protection essential (as height exceeds 55m)
6.	Structure like hydro power station sufficiently protected by means of surrounding structure e.g. high vertical cliff, high metallic structure or earth wire of transmission system (height 15m)	7	2	6	2	6	4	21	48	Protection can be omitted because the building is protected by surrounding.

2.6.2 Expected Direct Lightning (N_D) and Tolerable (N_C) Frequency Method: This is another assessment method for 100% need of protection, which is prevalent now-a-days in advance countries such as of Europe, UK, USA etc. Using French standard.

The risk assessment method takes into account the lightning risk due to following factor.

(i)   Building environment

(ii) Type of construction

(iii) Structure contents

(iv) Structure occupancy

(v)   Lightning stroke consequence

Taking into account of these factors the tolerable lightning frequency Nc are assessed and comparing it with expected direct lightning frequency Nd we can decide weather the lightning protection is needed or not.

2.6.2.1 *Expected frequency N_D:* The expected frequency Nd of direct lightning to structure is assessed by equation.

$$Nd = Ng \ max \ Ae \ C_l \times 10\text{-}6$$

Ng is yearly average lightning flash density in the region.

C_l is environment constant depending on building environment Nk.

2.6.2.2 *Evaluation of Ng max:* Nd or Ng max can be found by following supplied by manufacture.

1. Using strike density map N_a $N_g = N_a \ / \ 2.2$

2. Consorting lightning location network.

3. Using local isokeraunic level Nk. $N_g max = N_g \ / \ 2$

2.6.2.3 *Selection of C_l:* C_l is taken from the following table.

Sl. No.	Relative structure location	C1
1.	Structure locating within a space containing structure or trees of same height or taller.	0.25
2.	Structure surrounded by smaller structures.	0.5
3.	Isolated structure no structure with in a distance 3 times of height H	1.0
4.	Isolated structure on a hill top or a head land.	2.0

2.6.2.4 *Calculation of Ae:*

1. For a rectangular building of length L, width W & height H as shown in figure-1.
   $$Ae = L \times W + 6 H (L + W) + 9 \ \pi h^2$$

2. For rectangular prominent part $Ae = 9 \ \pi \ H^2$

3. Cylindrical building of height H shown in figure-1 $Ae = 9 \ \pi \ H^2$

4. Rectangular building with prominent part :- The prominent part whether rectangular or not, can be regarded as a cylindrical building. The prominent part can be situated any where w.r.t. roof, accordingly a circle of radios 3 x H height of prominent part can

be drown with centre at the place of prominent part w.r.t. roof / floor.

So the equivalent area of such building $Ae = 9 \pi H^2$

As the circle with radius 3 H covers the equivalent area of remaining part of building of height h as shown in figure if 3H > (L + 3h) where the L is length of the building.

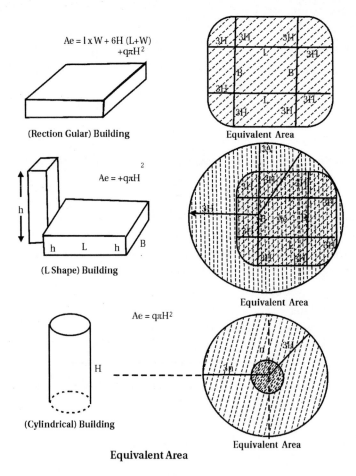

$Ae = l \times W + 6H (L+W)$
$+q\pi H^2$

(Rection Gular) Building

Equivalent Area

$Ae = +q\pi H^2$

(L Shape) Building

Equivalent Area

$Ae = q\pi H^2$

(Cylindrical) Building

Equivalent Area

Equivalent Area

Fig. 2.6.2.4

2.6.2.5 Calculation of tolerance lightning frequency Nc: It is assessed through analysis of the damage risk taken into account of following suitable factor on which its depends. Constants C_2, C_3, C_4, C_5 are assigned with them.

1. Type of construction C_2.
2. Structure contents C_3.
3. Structure occupancy C_4.
4. Lightning stroke consequences C_5.

These factors are selected from the following tables.

Table C₂ structure coefficient				
Structure	Roof	Metal	Common	Flammable
Metal	-	0.5	1	2
Common	-	1	1	2.5
Flammable	-	2	2.5	3

Table C₃ structure contents	
No value and non flammable	0.5
Standard value or normally flammable	1
High value or particularly flammable	2
Exceptional value irreplaceable or highly flammable, explosive	3

Table C₄ structure occupancy	
Unoccupied	0.5
Normally occupied	1
Difficult evacuation or risk of panic	3

Table C₅ lightning consequences	
Service continuity not required and no consequence on the environment	1
Service continuity required and no consequences on the environment	5
Consequences on the environment	10

Let $\quad C = C_2\, C_3\, C_4\, C_5$

Then $Nc = 5.5 \times 10^{-3}\,/C$

2.6.2.6 Protection needed or not: After computing Nc and Nd, we compare them to decide whether LPS is required or not.

So if Nd ≤ Nc LPS is not required

If Nd > Nc LPS must be installed with effectiveness E = 1 - Nc / Nd

The protection level is found by the value of E, in which range it lies as shoen below.

If E lies	Level	Peak Current I(kA)	Initiation distance D(m)
E > 0.98	1+Additional	-	-

$0.95 < E < 0.98$	I	2.8	20 m
$0.80 < E \leq 0.95$	II	9.5	40 m
$0 < E \leq 0.8$	III	14.7	60 m

This method also takes into account almost same parameters as in case of risk index method N_c, N_d are calculated after deciding the value of the constant C_1, C_2, C_3, C_4, C_5 from the table concerned with different parameters and A_c are found from height (H), length (L) and width (W). In risk index method L & W are not taken in account so this is a better method as compared with risk index method.

2.7 Zone of Protection:

The zone of protection of lightning conductor denotes the space with in which, it is to provide protection against a direct lightning stroke, by directing stroke to it self. According to best theoretical conditions a single vertical lightning conductor of any height attracts to it self all lightning discharge of average or grater intensity which in its absence would have struck a circular piece of ground round its base, radius of which equals to twice the height of conductor. Weaker discharges are attracted over shorted distances and the presence of other conducting objects with in space reduces protected area.

For adequate protection against all lightning discharge, the zone of protection of a single vertical lightning conductor is defined as cone with apex at highest point of conductor and with base of radius equal to the height of conductor and not twice its height. Similar consideration applies to horizontal lightning conductor.

So protecting angle of either horizontal or vertical lightning conductor is taken as 45°. Two or more vertical lightning conductor of equal height and spaced at distance not exceeding twice their height, the equivalent angle with in space bounded by conductor may be taken as 60° to the vertical. While away from conductor the angle is still taken as 45°. For two parallel horizontal conductors, the area bounded between them is better protected, provided no point of area to be protected is more than 9m away from horizontal conductor. Illustrations are shown in following figure-2.7.

Zone of protection with ESE lightning conductor developed in recent past, is The sphere of radius called protection radius. We shall discuss it later on.

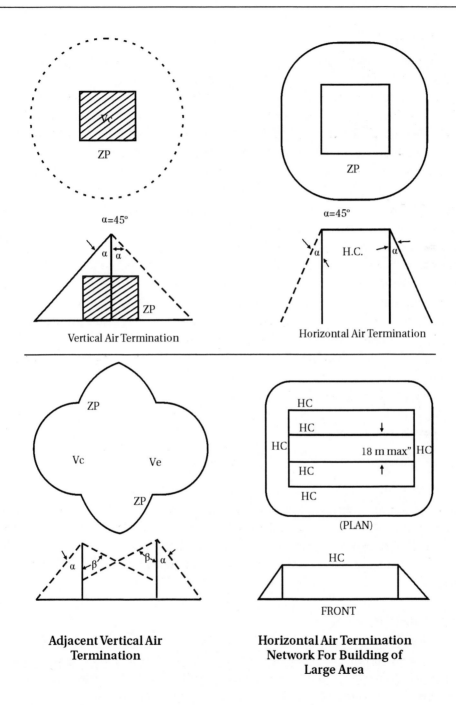

Vertical Air Termination

Horizontal Air Termination

Adjacent Vertical Air
Termination

Horizontal Air Termination
Network For Building of
Large Area

Fig 2.7

2.8 Design of LPS:

Lightning protection system must be such that all the lightning discharge must be discharged through it and the effective impedance to the passage of lightning current, between air terminal and earth must be least. Using more than one path between air terminal and earth the impedance further reduces. So, at least two down conductors are used for large buildings. The down conductor may be more. Design of LPS consists of design of its components namely air terminals and down conductor, after selecting proper air terminal.

Design of earth connection is equally important. Earth connections are evenly distributed. Further if there are metal objects of considerable size with in 2m of any lightning conductor, there will be tendency of spark or side flashes to jump from conductor to metal object. Damage due to this flash over is avoided by using interconnection or bond between conductor and metal object.

2.8.1 Necessary Information for Design: The following information must be obtained after examining the structure or its drawing if structure has not been built.

(i) Metal used in the roof, walls or reinforcement above or below ground (i.e. steel piling)

(ii) Available position of down conductor and air terminal to avoid external heating and spoil architectural or aesthetic beauty of building.

(iii) presence of non continuous or continuous metal object present on or in the structure.

(iv) The nature and resistivity of soil to design suitable earth connection.

(v) Presence of sound, radio or TV aerial.

(vi) Flag mast, plant room at roof, water tank and other salient feature.

(vii) The construction of roof for fixing of conductor.

(viii) Outline drawing incorporating forgoing details & main components.

2.9 Component Parts of LPS & Their Installation:

The principal components of lightning protection system are as below considering installation in two parts i.e. ELPI & ILPI.

1. ELPI (a) Air termination (b) Down conductor (c) Testing joints.

2. ILPI (a) Equipotential interconnection, joint or bonding (b) Earth terminal (c) Earth electrode.

2.10 Air Termination:

2.10.1 Verticacl & Horizontal Air Terminations: As discussed earlier air terminations are either vertical finial rod or horizontal strip, which are considered as equivalent and are selected purely on practical consideration. Vertical air termination is used on spire structure, while horizontal air termination on ridge of flat roof of small buildings or system of horizontal and vertical conductor for the protection of bigger buildings as to give sufficient zone of protection

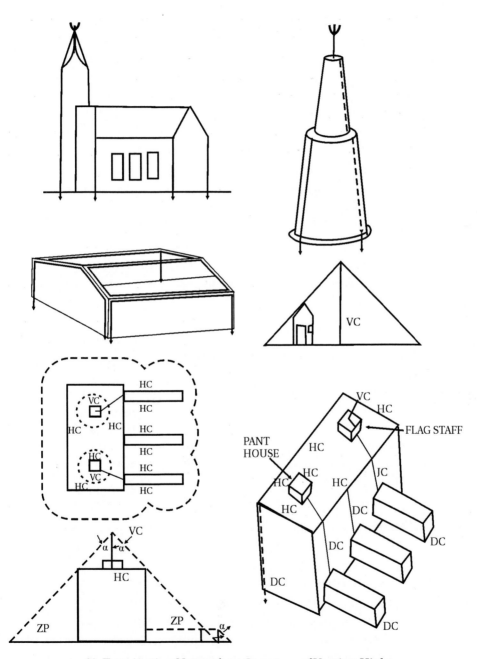

Air Termination Network on Structure of Varying Hight

FIGURE- 2.10

Only one vertical air termination is sufficient which shall be placed 30 cm above the spire object on which it is fixed. Horizontal air termination should be fixed on roof in a closed inter connected network, that no part of roof is more than 9m away from the nearest horizontal conductor except than an additional 30m may be allowed for each 30m horizontal conductor on top of the buildings to be protected, which is 30m below the nearest protective conductor which is not protected from the conductor. The horizontal air terminal is coursed along ridges parapet and edges of flat roof. Layout suits the shape of roof and architectural feature of building to cover every salient point.

All air terminals shall be fixed with the object to be protected by means of substantial braces. The fixing to building should be simple, solid with little effect of corrosion.

Some examples of protection of various types of buildings including, with large area of roof of varying profile, a thatched rondavel building with explosive hazard and a structure of varying height are illustrated in figure - 2.10.

Following type of protection of higher degree are recommended for building containing highly sensitive explosive or highly flammable material.

Type of building	Recommended type of protection
1- Building with explosive duster or flammable vapor risk.	Integrated mounted system with vertical air termination 1.5m high and horizontal air termination spaced 3m from each other (not 9m) depending upon (type of storage and processes) risk.
2- Explosive storage building and explosive work shop.	Integrated mounted system with vertical air termination 0.3m high and horizontal air termination spaced 7.5m.
3- Small explosive storage building.	Vertical pole type.
4- Building storing more dangerous type of explosive e.g. nitroglycerine (Ng) explosive manufacturing.	Suspended horizontal air termination at least 2m higher than structure with spacing of 3m.

2.10.2 ESE Air Termination: This air terminal gives a stronger ionization around the terminal through the use of electrical equipment independent of any external power source. This is really a effective system and requires to satisfy following conditions.

(1) ESE system need enough energy to generate a streamer when lightning occur independent of the climatic condition i.e. rain, sun or wind.

(2) The emission of streamer must be controlled by three following parameter.

(a) *Timing:* The emission of streamer must start immediately after lightning occurs, so the streamer from ESE meets or intercepts the down ward leader high surge form the cloud or down coming high voltage lightning surge to wreakers and to pass the weaker lightning surge currents safely to earth at an earth without causing damage, which may result if the streamer is delayed and lightning is allowed to strike or approach very nearer to the object.

(b) *Voltage:* For emission of streamer, high voltage spark is needed around the tip of the ESE system. This spark induces ionization phenomenon allowing the emission of a streamer. The ambient electric field is of the order of 10 KV/m before occurring of lightning round the downward electrode of ESE terminal (most common is prevectron 2) develops huge charge, which is stored in capacitor attached to downward electrodes, when the capacitor releases energy after sensing or detecting by a electronic device MCU.

A lightning discharge approaching ground, triggers spark between upper electrode and the central rod of the prevectorn, resulting the emission of upward streamer, which intercept the lightning approaching ground to give path to conduct lightning discharge vai down conductor   High voltage of the order 3500V generate streamer is controlled by a transformer connected between capacitor and upper electrode. The central rod of the ESE terminal is made of tin plated copper.

(c) *Intensity:* The intensity of spark necessary to generate emission of streamer is 20A. So this ESE system is better and does not allow strong lightning discharge to come down nearer to the object to be protected.

2.10.3 ESE Protection Area: To calculate the protection zone or ESE protection radius. Electrogeometrical module (EGM) of any air terminal is taking as rolling sphere. Air terminal may be vertical, horizontal, meshed cage or taut wire or ESE. The diameter of rolling sphere depends on lightning intensity I as $D = 10 \times I^{2/3}$.

The lightning strike point is any object on ground which may be flat surface or spear, the strike point is taken at a distance D from downward leader of lightning. The sphere comes in contact with vertical rod then the tip of rod is strike point as shown in figure.

It may be on ground surface, if it comes in contact with both single rod and ground at same time. There will two strike points as shown in figure but the lightning strike will not reach the surface between two strike points A and C.

This area since AO = OC so angle OAC = 45° so angle of protection for single rod ESE or any air terminal is 45° and radius of protection in case of ESE air terminal is calculated by

$rp = \sqrt{h\{(2D - h) + \Delta L (2D + \Delta L)\}}$

Where h = actual height of air terminal

ΔL = Path or length upward leader of streamer emitted.

$D = 10 \times I^{2/3}$.

D = 20m, 40m, 60m for protection level I,II,III

Now $\Delta L = V \Delta t$ where V = speed of lightning m / μs

Δt is the triggering advance time in micro second.

The software is also available for calculation of protection radius.

2.10.4 Installation and Function of ESE Termination: Every type of ESE lightning conductor have been designed, to conduct, the lightning discharge to the ground. After selection of proper ESE model, it is installed preferably on the highest point of the support with in the area that it protects. It may be machine room / gables or load bearing structure metal or masonry, chimney or mumty.

Further to measure the height of ESE terminal elevation mast of GI pipe can be used above the structure at highest point to keep the top of ESE position at 2m but may by more as per condition of site. If more than one ESE air terminal are used, they must be interconnected clearly by a conductor to down conductor. Indelec company on the basis of research and design, developed prevectron-2 ESE terminal. It is shown in figure 2.10.4.

When lightning is going to occur an intensive electric field is accommodated around the electrode so a huge amount of charge collected in a capacitor formed with electrode. This charge is not affected by any external condition i.e. rain, sun or wind.

An electronic device (micro controlled of same kind as PC) microprocessor unit or MCU connected to downward electrode detect lightning discharge approaching ground as it is programmed to read strong variation of electric field several MV/ ?s as downward leader potential 100 MV reached ground in 200 μs.

After detection it releases energy stored in capacitor to trigger sparke between upper electrode and central rod, resulting emission of an upward streamer which intercepts the lightning approaching ground to weeken it. Thus ESE terminal creates a conducting path between earth and cloud as central rod of terminal is connected to earth by down conductor and earth terminal enabling lightning discharge current to flow safely and continuously to earth from cloud.

2.10.5 Selection of ESE Terminal: The protection radius is calculated by equation.

$$R_p = \sqrt{\{h(2D - h) + \Delta L(2D + \Delta L)\}}$$

Where h = height of lightning conductor

D = 20m, 40m, 60m for protection level I, II or III

$D = 10 \ I^{2/3}$

I = peak current of first return stroke m KA

$\Delta L = V\Delta T$

V = velocity

ΔT = Triggering advance/μs

= $T_{SR} - T_{ESE}$

T_{SR} = mean trigger time of upward leader if a simple rod (S.R) lightning conductor (finial rod).

T_{ESE} = mean triggering time of upward leader of ESE lightning conductor.

Knowing ΔL, Rp from some graph (h, ΔL, Rp for different D protection level) supplied by manufacture drown or the basis of survey for a given (R$_p$) radius of protection needed at structure ΔL can be worked out so ΔT can be found by

$\Delta T = \Delta L / V$ for given velocity m/ μs, rigger time TSR for simple rod is fixed and know. Accordingly ESE terminal model is selected.

By experience and survey carried out by different model of ESE terminal. The manufacture has supplied some table showing R_p protection radius, carresponding to height of ESE for different protection level.

R_p for different model of ESE lightning conductor.

Level I, D = 20m

Height of ESE h(m)	S 6.60					
2	31	27	23	23	17	15
3	47	41	35	35	25	15
4	63	55	46	46	34	21
5	79	65	58	58	42	26
10	79	69	59	59	44	28

Level II D= 40m

2	39	34	30	30	23	15
3	58	30	45	45	34	22
4	78	60	60	60	46	30
5	97	66	75	75	57	38
10	99	88	77	77	61	42

Level D = 60m

2	43	38	33	33	26	17
3	64	57	50	50	39	26
4	85	76	67	67	52	34
5	107	95	84	84	65	43
10	109	98	87	87	69	49

Model	S6.60	S4.50	S3.40	TS 3.40	TS 2.25	TS2.10
ΔT	60	50	40	40	25	10

2.10.6 Advantage of ESE Terminal:

(1) The triggering time of corona effect is reduced by 70   s, as the low magnitude of lightning discharge or current passes through earth using ESE terminal instead of horizontal or vertical air terminal.

(2) Damage to appliance and material in side structures to be protected, is reduced due to reduction of increase potential difference due to lightning.

(3) In old system of mixed horizontal and vertical air terminal dense network of strips are laid on the roof which is avoided with ESE terminal to give less cutting or breakage of plaster of roof top to disturb aesthetic look and obstruction in movement on roof.

(4) By using one or two ESE terminal, we can protect long distance area as compared to old horizontal or vertical air terminal.

(5) The number of down conductors and earth terminal are reduced to one or two as per height of structure irrespective to plinth area of structure in using ESE terminal. So less cost is required using ESE terminal in place of old horizontal and vertical air terminal.

(6) Equipotential bonding are also reduced.

(7) The cost of installation with ESE terminal for big structures is less as compare to old horizontal and vertical air terminal.

(8) ESE terminal dose not require frequent maintenance and life is 25 years while old system has 9 to 10 yeas life and require frequent maintenance.

(9) Zone of protection or protection volume is much more as compared to old system.

(10)The ESE terminal has hemisphere zone of protection while old system has area of cone of 45° as zone of protection.

2.11 Down Conductor:

The down conductor is GI or copper strip of suitable size to pass the lightning discharge developed at air terminals to earth via earth terminal. Selection and design the number of down conductor is according to type of air terminals selected.

2.11.1 For Horizontal and Vertical Air Termination: The number and spacing of down conductor for vertical or horizontal air termination depend on the size and shape of building, architectural & aesthetical consideration. The number of down conductor is decided as per following consideration.

(1) The minimum number,. except for very small building is advisable to be two for a structure of base area up to 100m².

(2) The structure having base area more than 100m² one down conductor for 100m plus one for every additional 100m² or part there of or one for every 30m of parameter which ever is less.

(3) If the height of building is more than 30m as spire steel and flag posts following special consideration shall be applicable.

   (a) Non conductor structure of great height compared to length and width, in case of rectangular cross section or diameter in case of circular cross section, single down conductor is needed.

   (b) If non conductor chimney of over all width or diameter at top exceeds 1.5m at least two down conductor equally spaced and bounded by metal cap or conductor round the top of chimney must be used.

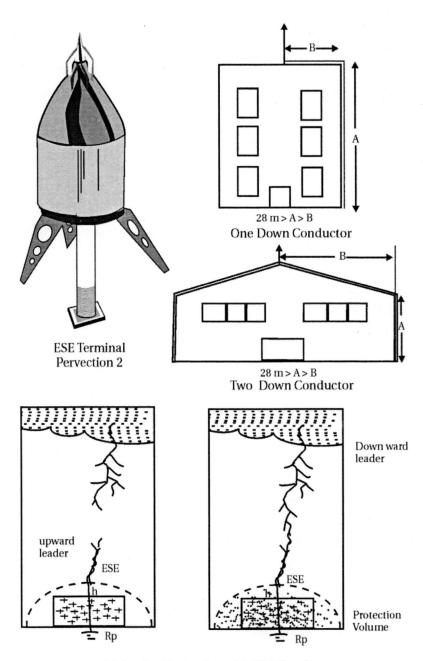

ESE Terminal
Pervection 2

28 m > A > B
One Down Conductor

28 m > A > B
Two Down Conductor

upward
leader
ESE
h
Rp

Down ward
leader

ESE
h

Protection
Volume
Rp

Lightning Protection with ESE Conductor

FIGURE-2.10.4

(c) Continuous metal frame or structure without external covering needs no down conductor and air terminals as well. Where more down conductors are needed two must be installed at more than 1.5m apart.

(d) Structure on bare rocks must be protected with at least two down conductors equally spaced.

Following precautions are taken for fixing down conductor.

(1) Down conductor should be distributed round the out side wall of the structure or building, so as to give direct path between air terminal and earth terminal. They run preferably along corners and other projection to avoid sharp bends, upturns and kinks.

(2) Lift well shall not be used for fixing down conductors.

(3) There must be no joints in down conductor unless it is necessary.

(4) Adequate protection is provided against the mechanical damage to down conductor.

(5) Down conductor must be so routed that it is accessible for inspection testing and maintenance.

Illustrational arrangements of down conductor with air termination are shown in figure -3.4. A special arrangement is made in case of following type of structure.

(1) A structure having high parapet, when the length of conductor forming loop at cornices or parapet exceeds 8 times of the width or the open side of loop parapet or cornices should be provided with holes through which down conductor can pass freely as sharp bend in down conductor produces high inductance voltage due to high impendence of entrant loop, and lightning discharge may jump across the side of top instead of passing through down conductor.

(2) In case of building cantilevered out from the first storey upward, the down conductor is fixed in the internal ducts rather than along contour of building around outer walls to avoid hazard creating to persons standing under over hang formed by cantilever.

2.11.2 Down Conductor for ESE Termination: Each ESE terminal connected with one down conductor in structure up to 28m. If the height is more two down conductor are, used with one air terminal. Down conductor is flat type strip made of copper steel of minimum 50sqm in cross section, so that high frequency impulse current circulates on the surface of conductor due to skin effect. Further if horizontal down ward projection of down conductor is larger than vertical, two or more down conductor are required.

Down conductor are placed on outer surface of two main walls such that its path is as direct as possible. All conductor or strips are fastened with clamps of same material or by solid rivet soldering or bracing. Drilling of down conductor should be avoided. A sleeve is installed on down conductor up to height of 2m from ground. Because of impracticable external routing full or part of down conductor may be routed inside inflammable service duct of building, though its effectiveness is reduced. When out side of building or structure has metal cladding or stone or glass curtain wall, the down conductor may be attached behind the cladding to concrete wall or load bearing structure. If cladding or supporting structure is conductive it must be bonded to the down conductor at the top and bottom end.

2.12 Joints:

The lightning protection system shall have minimum joints in it as far as possible. There must be no joints in down conductor below ground level i.e. in earth strip. The joints used in other part must be mechanically strong and electrically effective. In general all joints shall be welded, soldered or tinned except in case of bonds or test joints, where clamped bolted joints shall only be used. The joints with rod must be clamped or soldered type. Length of overlap in overlapping joints must be more than 20mm.

2.13 Test Clamp/Terminal/Joint:

Each down conductor should be provided a test joint/clamp used to disconnect the earth termination system for measuring earth resistance. Test clamp is usually installed in down conductor at 2m above ground level. When building has metal wall no, down conductor is used test clamps are inserted between each earth system and metal.

2.14 Bonds:

Any external metal, on or part of, structure is connected to the nearest lightning conductor air termination or down conductor by bonds, which are conductor of similar material as that of air termination or down conductor to which it is connected. These bonds have to discharge full lightning current.

If internal metal is to be bonded for equalizing potential, these bonds may be smaller in cross section than those of main lightning conductor as they carry small portion of lightning current.

All bonds shall be as shorts possible and similarly protected against corrosion.

In the reinforced concrete structure or frame of sufficient resistance to earth the reinforcement at top must also be bonded to air termination. Only those metals to outer surface which has not sufficient clearance or isolation are to be bonded. If the metal of sufficient length (e.g. cable, pipe, gutters, rain water pipe or stair ways) runs approximately parallel to a down conductor or bonds, it must also be bonded at each ends but not below test joints.

The gas pipe in no case be bonded with other metal parts, isolation which requires large clearance level to avoid side flashing is determined by

$$D = 0.3 R + H / 15n$$

Where
D = Required clearance in meter

R = Combined earth resistance of earth terminals in ohms

H = Height of structure in meters

n = Number of down conductor connected to common air terminal

In general isolation is practiced only in small building where it is difficult to obtain and maintain safe clearance.

2.15 Natural Components:

External, interconnected steel frame can be used as down conductor as they are conductive and their resistance is $0.01\,\Omega$ or less. In such case ESE terminal is directly connected to metal frame, whose lower end is to be connected to the earth terminal and equipotential bonding is provided. Natural component like internal metal structure, reinforcement, sunk metal structure sheet covering, to be protected or metal pipe and tank, if made of 2mm or more thick steel material, can be used to supplement down conductor.

2.16 Equipotential Bond:

As discussed under natural component like external and internal (sunk in wall) metal part equipotential bonding must be provided at its location in structure. Earth terminal, gas service pipe above the ground or underground must be interconnected with equapolential conductor similar as down conductor. Equpotential bonding for antenna or small post is done by using antenna mast spark gap type surge protective device with nearest down conductor. If water or gas pipe lines with insulated part with in the area to be protected must be bypassed by surge protective device.

For electrical or telecommunication system using screened conductor or conductors land inside metal conduit earthing, the conduit or screen provides sufficient protection. If not earthed (PVC pipe) active conductor must be bonded to the lightning protection system via surge protective device.

2.17 Lightning Flash Counter:

When lightning flash counter is provided, it should be installed on most direct down conductor above test clamp at height of about 2m above ground level.

2.18 Earth Terminal System:

Each down conductor shall have a independent earth terminal or earthing fixed at suitable damp location after propertesting. All the earth terminal should be interconnected below ground to common earth to equalize the voltage at various earth terminal to minimize the possibility of side flash over from one earth to other.

The resistance of earthing too effect the risk of side flash over with in structure where metal is present higher resistance may result in dangerous voltage drop in ground adjacent to earthing or earth termination GI or copper strip are used as earth terminal. The earthing with ESE terminal device is done in similar manner as with horizontal & vertical air terminal.

The material of down conductor & earth terminal must be same to avoid generation of thermo e.m.f. to oppose lightning current and increasing earth resistance. Earth terminal of least required resistance must be such that its horizontal & vertical component must not be excessive long.

The length of earth termination system depends on the resistance of soil in which earth system is to be installed. The resistance can be measured or known from standard table then, length of earth termination is given by L $= 2\rho$ / R   horizontal air termination

$$= \rho / R \text{ for vertical air termination}$$

R is desired resistance $\leq 5\Omega$ in plains & $< 8\ \Omega$ in hills

2.19 Earth Electrode:

Earth electrode must be buried and earthing is done as per IS 3043 of 1966 code. Salient points are enumerated below.

(a) Earth electrode shall consist of strips or plates of copper / galvanized iron (GI) of size 600 x 600 x 3mm.

(b) Earth electrode must be provided as close as practicable to down conductor depending on the neighborhood soil condition effecting the resistance e.g. dampness.

2.20 Fastener:

The lightning conductor shall be secured to the building or other object to be protected by fasteners. These must be substantial strong and made G.I. or other corrosion resistive material.

2.21 Inspection, Testing and Maintenance of Lightning Conductor:

2.21.1 Inspection: All the protective system must be examined by competent engineer after completion, alteration or extension to verify, that they are as per code IS 2300 of 1969. A periodical inspection at least once a year shall be done during maintenance.

2.21.2 Testing & Maintenance: Testing is done on completion or any alteration or extension of installation of protection system. Testing consist of checking of earth resistance, continuity of all conductor and the efficiency of bonds & joints.

A periodical testing is done even during maintenance once in two year normally, but in case of structure housing explosive and inflammable material testing is done in six months. The test results, condition of soil, its treatment must be recorded on the prescribed performa togather with particulars of engineers, contractors or owners of building responsible for the installation or maintenance or both, of lightning protection system.

If the resistance of the earth lightning protection system exceeds the lower value obtained at first installation by more than 100% their causes are identified and steps are taken for their removal and to maintain resistance.

For testing continuity of conductor and joints the ohm resistance of lightning protection system complete with end terminals but without earth termination should be measured and it must be fraction of a ohm or nearly zero.

Galvanization test of steel conductor may also be carried out, if needed. If G.I. shows sign of rusting conductor should be painted with suitable rust protective paint. If resistance increases, the conductor, must be replaced.

2.22 Estimating & Costing:

For the purpose of preparing the detailed estimated for finding cost of work we follow the following steps.

2.22.1 Requirement of L.P.S: It is essential to decide weather lightning protection system (L.P.S) is required or not by risk index method or Nc and Nd frequency method. However for certain building it is essential to use it because of security measure as discussed.

2.22.2 Selection of L.P.S.: Considering the importance aesthetic and cost of building, we select vertical / horizontal air terminal or E.S.E air terminal. Selection of vertical, horizontal or combination of both, are selected as per shape of roof of building. If the ceiling is peer shaped as discussed in 2.10,.the vertical air termination is fixed on the highest projected portion of spire building or tapered roof e.g. chimney, mumty or machine room etc. The horizontal air terminal GI / copper strip placed on flat roof/top at the edge or at 9m distance from each other, measuring on plan of the building or ceiling of building.

2.22.3 Design of Air Terminal:

2.22.3.1 Design of Vertical Air Terminal: The number of vertical air terminal can be found by counting number of vertical projection on the top of roof.

2.22.3.2 Design of Horizontal Air Terminal: The horizontal air terminal is placed at ridge of flat roof of building both lengthwise and widthwise, so length of horizontal air terminal is found by measuring periphery of roof of the building. If length and width more than 9m then strips are laid parallel to both length and width inside periphery and its length are also measured to give complete length of horizontal air terminal.

2.22.3.3 Design of E.S.E Terminal:

Considering condition of site so that the top of E.S.E terminal must be at heighest point of the surrounding object/structure, we decide the height of E.S.E terminal above the refernce structure roof including length of pipe mast and measure the distance of farthest point of structure to be protected as radius of protection, we refer the table provided by manufacture upto h<4m as per level of protection I, II, III On the basis of manufacture table we can find the type of model of E.S.E terminal and if radius do not cover the area of to be protected, more number of E.S.E terminals are used to cover complete area.

2.22.4 Design of Down Conductor: Down conductor for vertical or horizontal air terminal or E.S.E terminal design as per 2.11 of this chapter in the following manner.

2.22.4.1 For Vertical & Horizantal Conductor: Two down conductor is required for a small building of base area upto 100m ^{2} while building of base area more than 100m ^{2}. One down conductor for 100m ^{2} plus one additional down conductor for every 100m ^{2} or one for every 30m periphery of top of building, which ever less. It must be uniformly distributed over the periphery of roof. For the high building if height of the building is much greater than length & width or radius of building, only one down conductor is required. Similarly non conductive chimney of overall width or diameter more than 1.5 m, two down conductor are used.

GI or copper strip are normally used as down conductor. The distance of each down conductor from start point of air terminal to test joint is measured to give the length of down conductor required.

2.22.4.2 For ESE Terminal: Each ESE terminal is provided with one down conductor for building up to height 28m, if height is more, then two down conductors for each ESE terminal are used. Total length of down conductor is measured as in case vertical / horizontal air terminals and GI or copper flat strips are used.

2.22.5 Design of Bonds: Bonds to external and internal metal of building is connected to nearest down conductor or air terminal through bond. These are same strip as down conductor as discussed in 2.14. Total number & their lengths are counted/ measured to give total length of bonds or strip.

2.22.6 Design of Test Joints: As discussed in 2.12 of this chapter one test joint in every down conductor is made, so total number of test joint are counted.

2.22.7 Design of Earthing or Earth Terminal System: Every down conductor is provided with one plate earthing with strips as discussed is 2.18, 2.19. So the number of earth is counted and Length of strip is measured.

2.22.8 Detail of Measurment: After selection & design of LPS & its components we just measure or count the number of air terminal if vertical or ESE terminal number, of down conductor, test joint, bonds and length of strip (if horizontal air terminal) and bonds required for complete lightning protection system and summarize in table under details of measurement and totaled to find the quantities of these components.

2.22.9 Analysis of Rate: On the basis of market rates of each items as per manufacture price list of each item or components of the system required, we prepare proper analysis of rates to find the rates of the each items, if there exists no schedule of rates of these items. .

2.22.10 Bill of Quantity: Knowing the rates (scheduled or analyzed) and quantities of each items required for completion of installation of LPS, a bill of quantity is prepared on standard format taking each items in serial with their brief specifications. The total of bill of quantity will give the cost of installation.

2.22.11 Abstract of Cost: Abstract of cost is also prepared as per standard norms including the expenditure of miscellaneous items, called contingencies together with the cost of lump-sum provision of civil construction etc, if required and centage charges, if required. The total of the abstract of the cost will give the cost of the estimate or DPR etc.

3 | Fire Detection and Protection System

3.0 Introduction:

Fire may be caused in a building because of many reason, Flammable material wood, cotton or acrylic material, fabric made of cotton, wool or acrylic thread, curtain carpet or furniture etc. catch fire from lightened candle, kerosene lamp, bidi / cigarette or match stick and oven flame fire cracker or any spark due to short circuit in electrical appliances, machines or their wiring due to over loading or poor maintenance of electrical installation. Any material needs oxygen for burning so fire increases by getting oxygen from air.

The fire accidents in a building cause loss of building properties, damage of machines, equipments and appliance used in building, damage of costly records and may result in death of living being. These losses are not only serious but are irreparable. Thus for increased use and dependence on electrical equipment & appliances and other flammable material, it is essential to provide fire protection in building.

So any prevention or protection system against fire may save great losses as mentioned above. Fire protection is governed by an act enforced in every state.

In modern time the construction of multi-storied building is very common to accommodate the growing living requirement of mankind and decreasing available land for building construction. On fire accident, lift / shaft well or staircase works as chimney to create draught or vacuum to increase fire. So the multi-storied buildings above 10 stories are also termed as death well. So it has made mandatory to construct building or its architectural feature, as important route of lobby, staircase and passage in such a manner that it minimizes the possibility of causing fire and stop the spreading of fire, if it is created. It is well settled experience that people or living being die due to acute burning from fire or suffocation caused by smoke.

By fire act and building bye-laws, it has made mandatory to use protection against fire in a building. To provide fire, protection of fire its detection is important to know its nature, cause and place to take step to extinguish fire initially or as earlier as possible using pre-fire protection. Once fire increases we have to fight to extinguish fire by creating water cloud (much water) or fire retarding gas around fire i.e. post fire protection. In both detection & protection electrical energy has great role, so it has become part of electrical installation. Building manual or code of practice describes the provision of detection and protection system.

By getting detection signals we may communicate the out break of fire or fire accident to occupants by fire alarm or public address system to evacuate the building and to inform nearby fire brigade to reach the spot quickly to fight for early extinguishment to reduce the damage or losses. Occupants may use pre fire protection device often called as fire extinguisher.

So it is part of electrical installation in a building to provide fire extinguisher, fire detection & alarm system and fire fighting system.

3.1 Definition:

It is essential to understand various terms used in this chapter and to know their definition which are described as below.

3.1.1 Sections: The portion of premises or building, which is covered by one detection circuit at the indicating panel is called section. Division of section is done according to size and accessibility and shall not extend beyond a floor including mezzanine floor if any.

3.1.2 Fire Zone: The number of section on any one floor covered by one fire alarm circuit in a control panel comprises a fire zone. There may be one or more than one zone on a floor depending upon the size and accessibility.

3.1.3 Fire Sector: A sector comprises of a number of zone to indicate affected portion of zone in sector indicator panel.

3.1.4 Section Indicator Panel: This is provided in each zone to indicate affected section in the zone.

3.1.5 Control Panel: The control panel consists of both control and zonal indicator to indicate affected zone. So it is provided in a sector.

3.1.6 Sector Indicator Panel: It is provided in a remote manned center where an audio-visual alarm in control panel of each sector depicted and it indicates affected sector vai sector indicator. This type of panel is provided, where installation is divided into sectors.

3.1.7 Repeater Panel: A duplicate fire indicator panel to indicate the section, zone or sector only without control equipment is called repeater panel.

3.1.8 Trigger Device: The devices to initiate an fire alarm on breaking out of fire being operated automatically or manually is called trigger device.

3.1.9 Manual Call Point: It is a manual trigger device to give electrical signal to fire alarm provided in fire alarm circuit in the zone. It is also called manual call box.

3.1.10 Spot Indicator: A visual indicator provided externally at the top of the door of closed room or premises to indicate the triggering of detector with in room or premises.

3.1.11 Detector: A device to sense the source of out break of fire i.e. heat, smoke or flame and to give electrical signal automatically to the spot indicator & different panels.

3.1.12 Mimic Diagram: A graphics representation of the protected premise and their sub division or protection with indicating devices for each sub division or portion to indicate place of detector gives fire alarm to trace rapidly its sources in the premises.

3.1.13 Fire Alarm: The device which gives sound signal to alert the people about-break of fire and are activated by control panel e.g. gong bells, hooter, siren or horns etc.

3.1.14 Panel Sounder: It sounds with audible signal from control panel on operating of trigger device in any zone. It is provided in control panel.

3.1.15 Loop or Circuit: A loop or circuit of fire system is defined as circuit containing all trigger devices, indicator, MCPs, sounder, isolator etc. and fed from control panel.

3.1.16 Monitesred Wiring: The wiring used to warm or to give an alarm to indicate open circuit or short circuit failures.

3.1.17 Light Lantern: A translucent construction above the roof to admit light to the space below is called lantern light.

3.1.18 Fire Compartment: Portion of building separated by fire retardant (resisting) doors along the escape routes to prevent spread of fire and smoke.

3.1.19 Fire Extinguisher: This is a fire protection equipment used to extinguish fire during initial stage of out-breaking of fire as a result of burning of flammable material as cotton or acrylic cloth, rubber, plastic or flammable liquid like petrol, paints or gases and electrical equipment. It is called first aid fire protection equipment as it is useful only to extinguish fire at initial stage.

3.1.20 Fire Hydrant Valve: This is an outlet valve to stop or open the flow of water connected with a T shaped piece of 4" MS pipe to the riser or down comer or pressure pipe ring to use the pressurized water to extinguish fire with the help of hose pipe.

3.1.21 Stop Valve and Non Return Valve: Stop valve is an outlet valve to allow or stop the flow of water while the non return valve NRV allows to flow the water in one direction only and stop the return of water in opposite direction.

3.1.22 Fire Resistance Material: The material which withstand the fire for certain period are called fire resistance material.

3.1.23 Down Comer or Riser: Down comer is a MS pipe of 4" or more diameter use to flow of water at pressure from over head water storage tank to hydrant valve situated at different floor of a multi-storied building or ground floor hydrant ring, while riser is a 4" or more dia. MS pipe to carry water at pressure from fire pump at storage sump(ground water tank) to different hydrant valve.

3.1.24 Hydrant Ring: This is pipe line in the from of a ring around the building in which water at pressure remains with the help to fire pump to supply water to hydrant valve at ground floor.

3.1.25 Hose Pipe: This is canvass pipe of 100mm dia which when connected to hydrant valve with the help of coupling gives jet of water to extinguish fire.

3.1.26 Hose Reel: This is 19mm 30 meter swinging type rubber hose pipe in the from of reel when wrapped, used to spray pressurized water through a nozzle in the from of jet thrown to fire.

3.1.27 Water or Foam Sprinkler: This is a device which sprinkles water or foam when supplied to it with pressure, which rotates also to throw the water or foam.

3.2 Fire Protection System:

There are two branch of installation of fire dectection and protection system in building namely (a) Fire detection & alarm system (b) Fire protection or fighting system. The scope of this chapter is to know them, their design, installation testing, commissioning and its estimation & costing.

3.3 Fire Detection & Alarm System:

This is also called simply fire alarm system. The salient feature of the system is to sense, detect the place (section, zone or sector) of out-break of fire in the building or premise and communicate to the requisite persons to alert them about the out-breaking of fire, its location or origin or both in audio or video signal. Fire detection systems are crucial in saving life, protecting property and providing safety to industrial, commercial and residential buildings. The system comprises of various equipments and components installed in the buildings or premises to provide detection and to give alarm. The fire detection alarm system may be convenient or intelligent system and manual or automatic fire alarm system. The various equipments or components with, their brief specifications are mentioned in following sub-heads.

3.3.1 Manual Fire Alarm System (Moefa): Its design, installation & testing etc. are given by NBC code. The system is triggered by manual trigger device viz manual call point. The system is so designed that MCP near by located with the origin of fire out-break is actuated manually to operate panel sounder and other fire alarm sounder automatically or simultaneously to alert fire fighting or assistance staff, occupants and to take action to fight fire and to evacuate the premises.

The manual system consist of manual call points control panel, battery unit, fire alarm sounders and interconnecting wiring and cabling. Let us discuss them in detail.

3.3.1.1 Manual Call Point: This is also called manual call box. These are made of 1.5mm sheet enclosure with a frangible element on front, designed to be broken by impact or steady pressure to release push button to close the open electric circuit to actuate fire alarm at control panel. The front face of the MCP has area not less than 5000 mm² and its frangible element have exposed area not less than 1600 mm² in the shape of a rectangle or square. The frangible element is scratched by diamond bit so that it can be broken by finger pressure. MCPs are either surface mounted or recessed partly. All visible surface of MCPs except the frangible element is red painted and on front it is clearly written "in case of the fire break glass" with 5mm high letter.

MCPs are located at easily accessible place in premises e.g. lobbies on each floor at exit open on ground floor or any other convenient place to be approach-able by any person from any part of premises at distance not exceeding 22.5m. In hazardous area some additional MCPs are provided. The MCPs are fixed at the height of 1.4 m from the floor.

3.3.1.2 Control Panel: The manual fire alarm system has category C panel i.e. it does not identity any fault but gives only fire alarm in case of any fault. The false alarm may give confusion & panicky among occupant. Through this panel has equipments for reception, control, monitoring, recording and relaying of signal originating from MCPs to activate fire alarm sounder.

The control panel consists of zone visual indicator which is displayed on it identifying the location of zone of triggered MCP. The zone indicator may show location on numbered list of zone or plan of building mounted near indicator panel. A mimic diagram often form the part of indicator panel on the control panel.

The provision is also made in control panel to give fault warning in the event of following fault by at least an audio warning from panel sounder and a visual indicator, of affected equipment or sounder circuit or zone.

(i) Failure or disconnection or abnormal voltage of normal or stand by power supply.

(ii) Failure or disconnection of the battery charging equipment.

(iii) Removal of any MCP from its transmitter or power supply.

(iv) Short circuit or disconnection of the leads to MCP.

(v) Rupture or disconnection of any fuse or any protective device.

Control panel is provided with a sounder or sounders to sound a fire alarm on activation of any MCP in any zone. The panels also have sounders to give isolation warning on silencing of fire alarm.

All above sounders have different tones or one tone with suitably modulated for easy identification on panel. In case of silencing of fire alarm visual indication remains on panel till silencing fault is over & MCP is set right after replacement of its glass. Fire alarm sounder can be silenced by a switch too.

The control panel is made of 1.5mm thick steel cubical enclosure wall mounted type and antirust treated and suitably paintsed. The provision of sufficient size & number of glands of cable is provided in panel as required. A built-in 10W lamp is provided to illuminate the panel & building plan etc. Dual red lamp or LED to indicate fire condition and amber lamp or LED to indicate fault conditions with in zone. Pilot lamp for system ON, a.c. supply fuse condition indicator are also provided on panel. Visual indication for a.c. supply failure is provided in control panel. A reset push button for resetting of a.c. circuit in case of fire alarm condition. Every fire alarm sounders with any zone is connected to panel to evacuate the people. Fire control panel is situated in low fire risk area. It should be accommodated on ground floor near main entrance of building.

3.3.1.3 Fire Alarm Sounder: Fire alarm sounders of weather proof construction are used to give audible alarm for the evacuation of people from affected area. These sounder in each zone are installed on surface of the wall directly above or near MCPs. Additional sounders of high intensity are provided near the entrance in building at ground floor. The sound level of sounders in building is between 100db to 110 db and that at entrance between 125 db to 135 db at a distance of 2m from sounders. These sounders or its cage is painted red marked "fire alarm".

The fire alarms of any zone or zones are actuated by control panel. Panel is also provided a master switch to actuate all the sounder together. The panels actuate all the fire alarm sounder in a building automatically, if the key switch for silencing panel sounder, is not operated with in 5 minutes from the initiation of the alarm.

3.3.1.4 Battery Unit: A battery of adequate capacity to operate system for 24 hours with automatic dual rats A 220V 50hz single phase a.c. charger is provided with each control panel. The battery unit is housed in 1.5 mm thick steel cabinet which is painted after treating antirust protection. Battery panel or power unit panel have 0-30V voltmeter, 0-200 A ammeter, indicator lamp for mains & DC out put. The main supply is through 3 pine plug top with proper earth terminal to connect earth on current carrying metal part of system.

3.3.1.5 Wiring: 1.5 sq. mm PVC insulated Aluminum conductor cable in steel conduit is used. Both cable & conduit confirm relevant IS code to connect MCP. The wiring can be done in conduit either on surface or in concealed manner 2.5 sq. mm PVC insulated aluminum conductor wire for detector and 4.0 sq mm PVC aluminum conductor for sounder. PVC sheathed PVC insulated armored cable is used on surface in place of cable in conduct or to be laid in ground for making circuit or loop in premises to connect zones or sector. No joints are made in general but if it is unavoidable a suitable or accessible red painted junction box duly marked "fire alarm" is used to enclose joint. The maximum number of wire in a conduit is covered by PWD specification for easy drawing in conduit. Insulation test of conductor cable is performed before it is put to use. Equivalent size of copper conductor if used is selected in place of conductor.

Routine periodic test (daily, weekly, quarterly) are done for testing of equipment. Selection of number & type of equipment of alarm system is done as per design discussed later.

3.3.2 Automatic Fire Alarm System: In the automatic fire alarm system the trigger device known as fire detector, automatically trigger the fire alarm system and panel sounder to alert the staff to fight the fire & occupants to evacuates the premises or building where fire has broken out. The system quickly response to fire, detects it location and helbs in evacuating the premise quickly.

In this automatic fire alarm system except in small building or premises the extensive installation involving a huge building or a group of premises with common control, the entire building & area is divided into sector and each sector into zone. So a sector indicator are used. Each sector shall have a control panel and separate alarm circuit. Sector indicator panel is provided in remote manned center e.g. sub station care taker's room security office etc. Sector indicator panel will only indicate the staff about the affected zone and for exact location of fire the staff has to view control panel of the affected sector.

In this automatic system unlike manual system, the zones are further divided into sections which have section indicator panel with control panel to assist the staff to locate hall or room or exact point of origin of fire in the installation. Spot indicator is also used to indicate the affected detector of closed room. Section indicator panel are also provided, if needed separately in easily accessible places of the premises or group of zone, say stair case or lobby. Location of panel & division of zones are such that maximum distance to travel by fire fighting staff from control panel to fire affected place or search distance must not be more than 30m. Accessibility, size, provision of various detector is important to select the number of zone and floor area of huge building or distance of premises is considered to select number of zone or sector.

In modern automatically system manual call points MCP are in variably provided to enable the system to be triggered manually too in the event of a fire. Such automatically fire alarm system is called conventional fire alarm system.

The other automatic fire alarm system is analogue addressable fire alarm system which have all loop devices addressable to provide panel with unique identification of each device maintaining original sensibility of sensor enabling for alarm ventication over a time period day & night sensibility.

3.3.3 Conventional Fire Alarm System: This automatic system consist of following equipment. The specification and criterion of selection are described below.

3.3.3.1 Fire Detector: The device as explained before is to detect one, two or all three categories of fire i.e. smoke, heat or radiation (flame) and respond to anyone of these manifestation.

(A) Heat detector:

'A' Category:- These consist of thermister bimetal or a pneumatic tube. These work as thermal relay to trigger the device with less thermal delay. They operate at 57 °C± 1.5 °C tolerance. These confirm to IS 2175-1977. The automatic fire alarm system having 'A' category heat detector corresponds to notation AFAS-Th. They are used where outside temperature is less than 57 °C.

'B' Category:- It consist of a fuse link which absorb the heat of fusen to work as relay and has inherent thermal delay to trigger the device. These operate at 70°C ± 7°C tolerance. The automatic fire alarm system, having 'B' category heat detector corresponds to notation AFAS-Fl. These are used where ambient temperature exceed 43°C e.g. kitchen, boiler house, furnace room etc. These conform to IS 2175-1975.

Heat detector of category 'A' is often used. Hest detectors of both categories are of three type (a) fixed temperature detector (b) rate of rise detector (c) rate com pensation detector.

3.3.3.2 *Fixed Temperature Heat Detector:* This is a device which respond when its operating element becomes heated to a predetermined or fixed temperature. The sensing or operating element may be a bimetallic of deferent coefficient of expansion or a electrical resister with its resistance varying with temperature or a special eutectic metal which melts at fixed temperature or a liquid whose volume changing with temperature or line type element having two current carrying wire separated with heat sensitive insulation.

3.3.3.3 *Rate of Rise Detector:* It is a device which respond when temperature of the device rises at a predetermined amount. It is a line type detector having a small diameter copper tubing or spot type having diaphragm contact and compensating vent In an air chamber or a thermocouple or thermopile element.

3.3.3.4 *Rate Compensation Detector:* This device respond to fixed or predetermined temperature of air surrounding the device. This is a spot type with a metal tubular casing, which expand with rise of air temperature surrounding device.

(B) Smoke detector:

As the name suggest this type of detector trigger respond quickly invisible smoke from a clear burning fire but respond slowly to optically dense smoke where there are production processes which produce smoke to operate smoke detector. These are ideal for use in dust free and humidity controlled atmosphere as deposit of dust / moisture on the sensing element of detector causes fault or unwanted alarm, so these are not used in damp and excessive humid area as in industrial area and battery room, where combustion product like fire gases fumes vapor are present to give false alarm. These are not used in cold (below 0°C) or very hot equipment (above 50°C). Alternative should be used to smoke detector. There is no Indian standard but foreign standard selecting smoke detector on ceiling height, shape and surface ventilator etc.

There are four type of smoke detector.

3.3.3.5 *Ionization Smake Detector:* This is most expansive and most sensitive detector and gives an early response to fast burning Fires caused by cotton or paper etc. This features relative sensing of both visible and invisible products of combustion. They have advanced dual chamber single radioactive source designed to operate on radiation of inside flame or scattered light by smoke. Its chamber arrangement automatically compeusate for environment change like humidity, Ambient temperature &

atmospheric pressure. These smoke detector are available in different sensible version.

3.3.3.6 Optical Smoke Detector: These are less sensitive and operates on the falling light scatter principle due to solid particle in smoke on photo sensitive device. So it is effective to provide an early response to wide range slow burning smoldering fires. It does not respond to fires from petrol alcohal etc. These are also available in different sensitive version.

In other type photoelectric or optical gas smoke detector the photoelectric lights normally falling on photo sensible device is obstructed to actuate detector.

3.3.3.7 Semiconductor Smoke Detector: These detector respond to absorption of some combustion products in particular, causing oxidation of gases by electrical change in, semiconductor, photoelectric device. These are rarely used.

3.3.3.8 Resistance Bridge Smoke Detector: This device utilise resistance bridge principle. It responds to fall of combination of smoke particle & moisture on conductoring electrical bridge grid to increase conductance of grid & actuate detector.

(C) Flame detector:

These detector respond to visible radiant energy to human (4000°A to 7000°A) or outside the visible range above 7000°A i.e. ulttra-violet infra-red radiation. These device has photoelectric cell sensing element which actuates when exposed to redlaut energy of flickering of flame. These detectors are not used.

3.3.3.9 Response Indicator: Response indicator is provided to indicate the actuating or triggering device connected to it. So it is used where room / space are closed. R.I. are placed out side the room or place above the entrance door of these room / space. So this will indicate fire accident initiated in this room or space.

3.3.3.10 Manual Call Point: These are also used in convention automatic fire alarm system as described in manual fire alarm system.

3.3.3.11 Fire Alarm Sounder: These fire alarm sounder are provided as described with specification in manual fire alarm system.

3.3.3.12 Sector Indicator Panel: These panel located in remote manned center consist of equipment for reception and recording the audio-visual signal for both fault and fire alarm condition originating from control panel provided in each sector. This panel also gives fault warning due to open-circuit or short circuit fault in interconnecting cable between sector panels and the various control panels connected to it. Sector panel also detects any components and connection failure with in panel and also in control panel via fault warning indicator and sounder called sector indicator and sector panel sounder respectively.

Sector indicators on sector indicator panel identifies number of sector by sector numbered list or plan of premises or building mounted near indicator panel, or it may identify the sector on a mimic panel, if provided with sector panel. Sector panel also have fault warning audio indicator & sounders and video indicator if provided with PC monitor. Sector indicator panel may have a sounder or sounders to indicate fire alarm, fault warning or isolation warning of a faulty circuit with one or more tones with provision of silencing key switch.

The well mounted type selector indicator panel is enclosed in 1.5mm thick steel sheet dust & vermin proof cubical enclosure with anti rust treatment and duly painted. They comprises adequate size & number of cable gland for cable entry. A built-in 10W lamp to illuminate the panel and building plan, if provided on panel. Dual red lamps and Amber lamp or LEDS for fire and fault signal respectively, pilot lamp for system ON. All lamps are fed from panel power supply unit. These panels also have AC fuse indicator (MCBs) reset push button for resetting of the electrically locked circuit in case of fire alarm condition and sector panel sounder and visual indicator of AC failure. The sector indicator panels also have tests facilities for simulating fire short & open circuit fault condition with sector isolation facilities.

3.3.3.13 Control Panel: Conventional system use control panel of category A which can identify open circuit, short circuit, earth fault, removal and failure of detector or MCP (trigger device) component or connection failure as a fault and provided a fault warning and indication with fire alarm or indication and its origin in a zone. Location of zone may be shown by zone numbered list or plan of building mounted near the indicator panel or in mimic diagram if provided with control panel.

The control panel is capable to give both audio & visual warning of the affected equipment or sounder circuit or the zone. It must also give fault warning in the event of following.

(a) Abnormal voltage or failure or disconnection of the normal or stand by power supply or battery charging equipment.
(b) Removal or failure of any trigger device i.e. detector or MCP if used.
(c) Short circuit or disconnection of leads to trigger device or alarm sounder.
(d) Rupture or disconnection of any fuse or operation of any protection device to prevent to give fire alarm.
(e) Removl of any component in control panel e.g. relay, circuit & sounder etc.

The control panels have sounder or more sounders with one or more tones with facility to silence sounder by key switch. Control panel is enclosed in 1.5 thick steel cubical dust and vermin proof enclosure. It is wall mounting type. The enclosure must be anti rust treated and suitably painted to the desired finish. Different model of control panel provide control & indication facility for 2, 4, 6 or more zones compatible with many conventional device.

The control panels too have adequate size & number of cable gland for cable entry a built-in 10W lamp. Dual red & Amber lamp or LEDs to be fed from panel supply to illuminate panel & building plan to indicator. The control panel sounder and reset push button for resetting of relevant circuit are electrically locked by fire alarm signal. The control panel must have facility for zone isolation and test facilities for simulation fire and short or open circuit fault. The control panel is capable of connection to any additional fire alarm sounder.

The control panel and battery unit is accommodated in area of low risk and protected by unauthorized interference and fire by suitable detectors. The control panel is provided near entrance of the building ground floor for easy access by staff of security or fire brigade. A repeater panel can be used for any staff situated far from control panel and building. A control panel in a building with multiple occupations is located in a area that is in common use by all occupant.

A control panel must have operational instruction in the event of fire adjacent to it and a log book recording inspection of the system fire, or fault incidents with causes and action taken is kept near panel. A control panel is never put out side building or in any room of severe environmental condition.

3.3.3.14 Section Indicator Panel: If zone is divided into sections section indicator panel at any prominent accessible place & comprise of equipment for reception and recording if the signal originating from trigger device and releasing such signal to control panel, section indicator panel like sector panel have section numbered list, building plan mounted near the panel or it may identified section on mimic panel if provided.

Section indicator panel like sector indicator panel is wall mounted type and enclosed in a 1.5mm thick steel or CRCA sheet dust & vermin proof enclosure duly anti rust treated and painted to desired finish.

It also have adequate cable gland, a built-in 10W lamp. Dual red or Amber lamp or LEDs, pilot, AC fuse indicator, test facilities, reset facilities etc. like sector panel.

3.3.3.15 Wiring: Fire retardant or fire retardant low smoke (FR & FRL) PVC insulated copper conductor cable in steel conduct is used confirming to relevant IS code. The wiring is done either in conduit on surface or concealed 1.0 sq mm copper cable is used for MCP and A class heat detector, 1.5 sq mm copper for B class heat detector and 2.5 sq mm for sounder are used. Armored cable on surface are used in place of wire in conduct or to be laid in ground for connecting different zone or sector. No joints are made in general in wiring but if it is in unavoidable, a suitable accessible junction box with red paint duly anti rust treated and marked "fire alarm" is used to enclose joints. Insulation test of cable is performed prior to be used after drawing in conduct. The number of wire in conduct are down as per IS code & PWD specification.

Routine periodic (daily, weekly, monthly, quarterly) test are done for testing of the equipment & system.

3.3.3.16 Battery Unit: The battery unit is provided similarly as provided in manual fire alarm system.

3.3.3.17 A.C Duct Sensor or Duct Probe Unit: It is provided to monitor the air flow in ventilation duct, to shut down of the AHU in case of fire, stopping smoke from spreading. It is provided duct opening in both conventional and addressable fire alarm system.

3.3.4 Analogue & Addressable Fire Alarm: These are fire alarm system, which uses all addressable loop device i.e. addressable detector or sensors, call points, sounder, line controller & isolator and addressable analogue fire panel instead of conventional equipment to provide the panel with unique identification. This system quickly & reliably detect fire which is most important attribute of a good fire alarm system to save life & property. Each device maintain original sensitivity of sensors enabling fire alarm ventication over a time period. Addressable fire alarm panel are available for 1,2,4,8 and more loop with 100 or more device per loop.

Additional equipment provided with addressable fire alarm system have following specification and description.

3.3.4.1 Multi Sensing Fire Detector: This is a device which respond to both heat and smoke, the two combustion product of fire. On triggering these sensors give signal to spot indicator and sounder / isolator or control module through control panel. These are provided in similar manner as heat or smoke detector. These are provided when ambient temperature or other condition are at the border to choose heat or smoke detector / sensor.

3.3.4.2 Fault Isolation Module: These modules are provided to electrically isolate different section of detector circuit or loop in order to isolate short circuit. For cable damage, earth leakage, double address or other problem in loop. So it is provided in every loop.

3.3.4.3 Control Module: This module monitor open or short circuit in the wiring to the sounder as well as external power supply to the sounders. It must give assigned point address to the microprocessor & control panel. This is also provided in each loop.

3.3.4.4 Central Processor Unit.: C.P.U. of 240V 50Hz. 1 phase AC supply with 24V battery back up, is provided for video display on computer monitor and it will also communicate with printer and various microprocessor based control panel in addition to computer monitor. Indication and information received from detector or other field devices, switches etc to CPU, will be repeated to other panel, if desired.

3.3.4.5 Addressable Fire Alarm Control Panel (C.P): This panel has microprocessor based loop controlled utilizing technique. Such as early detection, safe test, automatic calibration, planned maintenance, wide range sensitivity, adjustment, alarm ventication and day-night control. The power supply to fire alarm control panel is 240 V 50 Hz single phase AC supply.

3.3.5 Advantage of Analogue Fire Alarm System: There are following advantage of analogue fire alarm system over conventional fire alarm system.

(i) *Early detection of fire:* Early detection of fire through active analogue system when all devices in the system individually addressed to provide the panel with absolute identification of each device, is possible.

(ii) *Safe test:* System verifies all the aspect of each sensors operation and communication including calibration & sensitivity to smoke. The system comes out a safe test automatically once every 24 hours.

(iii) *Automatic Calibration:* System monitors any changes caused by contamination and then adjusts the calibration accordingly to maintain each sensor of its original sensitivity.

(iv) *Planned Maintenance:* System checks the contamination level of every sensor and gives a service needed signal where cleaning is required. A service report permits maintenance to be done on at planned basis.

(V) *Wide Range Sensitively Adjustment:* Each sensor sensitivity is indivisually adjustable over a wide range either manually at the control panel or automatically on time zone basis.

(vi) *Alarm Ventication:* Each sensor in the system can be individually enabled for alarm ventication over a time period.

(vii) *Day & Night Control:* Provides separate sensitivity for day and night.

Analogue addressable fire alarm system is micro computer based loop controlled utilizing distributed technique. The system generally include central processing unit CPU power supply, fault isolation module, control module, multi sensing device, video display unit etc. complete in addition to addressable indicating device or detection device e.g. fire alarm panel, heat sensor, smoke sensor, loop sounder, MCP. Other features of equipment are same as in conventional system.

3.3.6 Selection of System: Some of example of application of different type of fire alarm system is given as below. These are guide lines in general, fire alarm system selection is done as per design and physical aspect. Where MOFFA is used with AFAS, it means MCP are used as additional equipment.

Type of building	Recommended			
	MOEFA	AFASF	AFAS Th	AFAS Sm
1- **Hostel & Residential** above 2 stories with floor area more than 200 sqm.	MOEFA	-	-	-
2- **School & College** above 2 stories floor area more than 500 sqm.	MOEFA	AFASFL	-	-
3- **Hospital** (a) Ward or duty (b)O.T & ICU (Air conditioned)	MOEFA	-	-	-
(i)Below false ceiling 2 in return air duct (ii) Above false ceiling	MOEFA			AFAS Sm
(c)X-Ray dark room (d) Rest hospital	MOEFA		AFAS Th	
	MOEFA		AFAS Th	
	MOEFA	AFASFL	-	-
4- **Office area** (a) **Ordinary office** above 2 stories with floor area more than 500 aqm	MOEFA	AFASFL	-	-
(b) **Telex room**	-	-	-	AFAS Sm
(c) **Monuments or important building** (d) **All centrally air conditioning area**	MOEFA	AFASFL	AFAS Th	-
(i) Below false ceiling (ii) Above false ceiling	-	-	AFAS Th	-
	-	-	-	AFAS Sm
5- **Computer installation** (centrally air conditioning) (i)**below false ceiling** (ii)**above false ceiling**	-	-	-	-
(i) below false ceiling			AFAD Th	
(ii) above false ceiling				AFAS Sm
6- **Technical area** (i) Libraries, Museum, Art gallery, **communication building**	MOEFA	-	-	AFAS Am

(ii)Laboratories terminal building	MOEFA	AFASFL	-	-
(iii) **All centrally air conditioned**	MOEFA			
(a)Below false ceiling	MOEFA (duct)			
(b)Above false ceiling	-	-	-	AFAS Sm
(iv)**Substation, tower, plants,** laundries, large work shop, cold storage AC plant room, pumping station, kitchen (non domestic)	-	AFASFL	-	-
7-Auditiorium (a) Non air conditioned.	MOEFA	- AFASFL	-	
(b)**Centraly air conditioned** 1-Below false ceiling	MOEFA (duct)	-	AFAS Th	-
2-Above false ceiling	-	-	-	AFAS Sm
8- Storage area Baggage room,	MOEFA	-	-	AFAS Sm
record room, Po	-	-	-	AFAS Sm

3.3.7 Design of Fire Alarm System: We study the plan of building provided for estimating and costing for fire alarm system. Knowing the feature of building and keeping the guidelines given vide 3.3.6, we select the type of fire alarm system and proceed for following next step.

The protected area should be divided into easily accessible zones, to give precise indication of that part from which the alarm is originated quickly. A zone is limited with in a fire compartment consisting of main walls, floor & ceiling of the building of floor area not exceeding 2000m². A zone normally cover only one storey except light wells, lift shaft, stair case or other flue like structure extending beyond one storey. If total floor area is less than 2000m² say 300m² then zone must not exceed more than one storey.

In large installation involving many premises under common control or a huge building or many adjacent small building, we divide all premises into sector which in turn are divided into zones with a separate sector indicating panel to identify each zones with separate alarm circuit or loop.

We may or may not divide zone into section to assist fire fighting or staff to locate the room / hall or exact point of out break of the fire with least delay.

Each section is fed by one loop / circuit of control panel. Each loop may have fixed number of device i.e. triggering device, response indicator, fire alarm sounder, controller etc. The number of these device in a loop is decided as per type of fire alarm control panel (FACP) selected. The circuit are two type for different functional / partition circuit (a) Detection circuit connecting detector in a section to C.P (b) Fire alarm circuit connecting sounder to C.P. These two are of different size as discussed earlier.

To prepare the detailed estimate be design for each equipment or component of fire alarm system.

3.3.7.1 Marking of Plan: Building plan is needed for all the building in the premises to be provided the fire alarm system. Trigger device like manual call point MCP, heat / smoke detector, response indicator, sounder, isolator etc. are to be marked on the plan of building as per design criterion depending on the floor area, height & size of room. Spacing between detector, existing of specially structural feature like false ceiling beam chauffer etc.

3.3.7.2 Design of MCPs: MCPs are provided at required placed as mentioned in 3.3.1.1 and marked on plane of the building as keeping the followings points in mind.

(i) These are provided in staircase, lobbies of each floor as well as at 2x4 to the abeu on ground floor. Additional MCPs are provided in hazardous place also.

(ii) Location of MCPs are such that a person from any point in building has not to travel a distance more than 22.5mm.

(iii) MCPs are fixed at a height of 1.4m above floor.

MCPs are generally provided with MOEFA manual fire alarm system, but as discussed earlier. Automatic fire alarm system also uses it.

3.3.7.3 Selection of Detector: Selection of detector in practice, as fire produces both heat and smoke, is done as per following purpose, but their design is as mentioned in above paragraphs 3.3.3.1.

(i) Fixed Temperature Heat Detector:- It is used where ambient temperature are high and / or rise & fall rapidly e.g. dust & humid areas with variation of ambient temperature.

(ii) Rate Of Rise Heat Detector:- It is suitable, where ambient temperature is low or sub zero and / or may not vary rapidly e.g. cold storage, huge deep freeze unit.

(iii) Optical Smoke Detector:- It respond to optical smoke not to invisible smoke so it is used only in dust free and clean atmosphere as dust or dirt on surfaces of hot sensitive element or lamp impair efficiency of detector.

(iv) Ionization Chamber Smoke Detector:- It is used in invisible smoke not in optical dense smoke, so it is used in dust fire and humidity controlled atmosphere.

Conventional Fire Alarm System

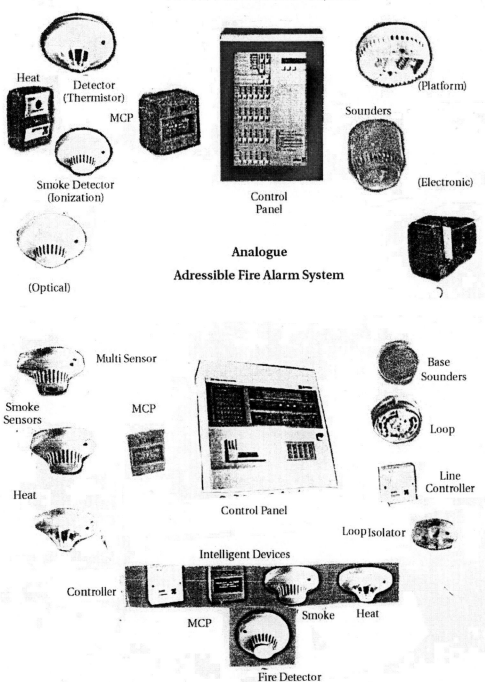

Heat

Detector
(Thermistor)

MCP

Smoke Detector
(Ionization)

(Optical)

Control
Panel

(Platform)

Sounders

(Electronic)

Analogue
Adressible Fire Alarm System

Multi Sensor

Smoke
Sensors

MCP

Heat

Control Panel

Base
Sounders

Loop

Line
Controller

Loop Isolator

Intelligent Devices

Controller

MCP

Smoke

Heat

Fire Detector

(v) Multipurpose Detector:- It is used when atmospheric condition are in midway or at the border to chose either heat or smoke detector.

3.3.7.4 Design of Heat Detector: Category and type A heat detector is selected for different location, spaces or portion of building to be protected as per their specification.

The spacing of heat detector A category for flat ceiling must be as per detailed in the following table.

Location	Area coverage	Maximum distance between center any direction m		Maximum distance from walls or partition	
	m²	m		m	
		Ordinary area	Corridor	Ordinary area	Corridor
1-Mounting height above floor up to 3m	46	7	10	3	6
2- Ceiling height more than 3m	37	6	9.5	3	6

The spacing of heat detector category A in concealed spaces (not more than 2m in height) must be detailed in following table.

Percentage of depth of beam to vertical height of con space	Maximum area covered by one detector	Maximum spacing between detector	Maximum distance between any wall or partition to nearest detector
1	2	3	4
50% or less	92 sqm	9m	6m
50% to 75%	46 sqm	7m	3m

Above 75% of depth each area is individually protected and concealed space protected as per following norms.

(1) Where protection area is segmented by ribbed ceiling, beam, joist or ducts with vertical depth is in excess of 25 Cm. such segment has at least one detector, if the area of segment is in excess of 37 sqm where ceiling is divided into smaller area then at least one detector is installed for a number of segments with area 28 sqm.

(2) If the height of ceiling exceeds 3m spacing distance give in table for flat ceiling is reduced 20%.

If the structure has a maniter saw tooth gable or lantern have a row of detector with in 60 cm from apex of roof. If the ceiling floor is surmounted by open joist, detectors are placed between joist at highest point. In the storage halls having long racks and shelves, the detectors are placed between shelves at 1.5m distance.

In the case of sectioned partition room or storage rack (exactly up to 30 cm from ceiling), each section must have detector. Similarly flue like opening & stair case must have detector at the top or on each floor respectively. The heat detector line or spot is ineffective over a height 9m.It must be positioned not less than 25mm & not more than 150mm below ceiling. One detector is placed at least on each floor with in 1.5m of height in elevators, stair case, well holes etc.

3.3.7.5 *Design of Smoke Detector:* The smoke detector are designed and selected as per following requirement, w.r.t. shape, surface, height of ceiling & ventilator.

(i) One smoke detector is provided for each 92 sqm (in UP and Delhi 60/70 sqm).

(ii) The spacing between smoke detector is not to exceed 12 meter in ordinary area and 18m in corridors.

(iii)The distance from the center of smoke detector to any wall or partition dose not exceed 6 meter in ordinary area and 9 meter in corridor.

(iv)In ceiling construction, if beam is 150mm or less in depth, it is taken as smooth or flat ceiling.

(v) If beams are more than 150mm in depth but less than 10% of ceiling height, the spacing of smoke detector in direction perpendicular to beam is reduced by tarce the depth of beam.

(vi) If the depth of beam exceeds 450mm or 10% of the ceiling height and are more than 2.5 meter on centers each bay (partition) is treated as a separate area.

(vii) If partition or storage rack extends up ward to with in 20 cm of the ceiling each section of such partition is treated as separate room or area.

(viii)In case of sloped ceiling the smoke detector is located with in one meter of the peak of the ceiling.

(ix)The smoke detector is placed to favor the air flow toward return opening. One additional smoke detector may be provided if needed due to other constraints.

(x) The smoke detector is used at places, where roofs or ceiling is of unusual shape, or have deep beam, in staircase, canteen, restaurant, plant room or in lofty building of 10m to 30m in height.

(xi) The smoke detector is used in an air cooled or air conditioner space or places of over head heating.

3.3.7.6 *Design of Spot or Responde Indicator:* Room, space or premises which is closed to enter if the they are not under use one has fire detector inside, spot indicator are provided out side at the doors to indicate the triggering of detector inside the in doors if every closed room / hall opened out side or-in corridor or lobby etc. is provided with this response or spot indicator. Door or doors of closed room opened in corridor /

lobby must have one R.I for more detector inside closed room, so one may get indication of out break of fire if he passes outside that room from any side. So numbers of response indicator are marked on the plan of the building accordingly.

3.3.7.7 Design of sounder or Hooter: Fire alarm sounder is to be given to alarm and alert the occupant. So it is situated at lintel level, .one above MCPs situated in lobby, staircase or corridor, one of high intensity at entrance on ground floor and one at the place where control panel is located, in similar manner in both type of fire alarm system.

3.3.7.8 Design of Control Panel: One control panel of suitable loops is provided to indicate zone in every zone of sector, if the premises is divided into many zones, according to type of the fire alarm system.

3.3.7.9 Design of Section Indicator Panel: If a zone is divided in to section a section indicator panel must also be provided, according to type of the fire alarm system.

3.3.7.10 Design of Sector Panel: One sector panel is provided to indicate sector if premises is divided into sector, according to type of the fire alarm system.

3.3.7.11 Design of Isolator or Isolator Module: This is used to isolate short circuit fault, cable damage earth leakage fault etc. in a loop or different section of a detection circuit or zone. It is provided in every loop.

3.3.7.12 Design of Controller or Control Module: This is provided to monitor open short circuit in the wiring to sounder as well as external power supply to the sounders. This is responsible for operating of sounder. This is also provided with each loop of control panel in case of conventional or addressable fire alarm system as discussed earlier.

3.3.7.13 Desifn of Mimic Diagram: One mimic diagram for indication of exact location of triggered detector or exact origin of out breaking of fire in the building. This is provided with every control panel.

3.3.7.14 Design of Repeater Panel: A repeater panel with each main control panel is provided at distant place where fire brigade staff or other security staff is available

3.3.7.15 Design of Wirinf & Cable Earthing: As discussed earlier both in manual or (conventional and addressable type) automatic fire alarm system (3.3.1.5 & 3.3.3.15) we use wiring to connect triggering device, fire alarm sounder etc. to control panel or other indicator panel. This wiring may be sub main wiring in concealed conduit or conduit on surface or cable on surface or laid in ground supported by earthing as per wiring code of practice or Indian electricity rules 1956 & Indian electricity act 2003 and PWD specification. In this installation current flows in only on triggering of triggering device to actuate control panel and associated sounder, indicator panel, mimic diagram etc. so plate earthing (G.I or copper) with one G.I wire is selected to earth control panel. Earth wire run along with wiring as per specification of sub main wiring. Sub mains or cable required can be measured as per plan of premises or building or as per actual on site of installation.

3.3.7.16 Design of Battery Unit: A battery for each panel is provided both in manual and conventional automatic fire alarm system, while 24 V back up battery is needed with AC supply mains in addressable fire alarm system. A 240 V 50 Hz single phase AC battery charger with three pin top is provided with each battery unit.

3.3.7.17 Design of PC Monitor: Suitable CPU, Printer, PC monitor is provided for video indicator or recording of line or fire fault.

3.3.7.18 Public Address System: Public address system can be used for addressing the people for alarming them assisting in the rescue operation.

3.3.8 Detaik of Measurtment or Schedule of Material: The number of MCPs, detector, response indicator, fire alarm sounder, control panel, section or sector panel, mimic panel or repeater panel according to type of the alarm system. are tabulated in the following table called schedule of measurement and exact number is found by summing of each Coolum equipment done.

| Sl.No | Location | MCP | R.I | Heat detector | | Smoke detector | | Multisensing detector | Sounder | Control panel | Mimic diagram | Controller | Isolator |
				Fixed temperature	Rate of Rise	Optical	Ionisition chamber						
1.	2	3	4	5	6	7	8	9	10	11	12	13	14
1.													
2.													

3.3.9 Analysis / Schedule of Rates: The rates of each item is either taken from the schedule of rates if it exist, other wise proper analysis is prepared on the basis of market rate to evaluate rate of the each item to be provided.

3.3.10 Bill of Quantity: Bill of quantity is prepaid on the standard format to find out the exact cost of installation of fire alarm system.

Sl No.	Type	Agent	Capacity	Effective Discharge	Jet Length	Discharge Time	Test pressure	Gas cartridge/ Storage pressure	Anti corrosive Substance
1	2	3	4	5	6	7	8	9	10
PORTABLE									
1	Water CO2	Water	9 ltr.	95% min	6m min	60s-120s	30Kgf/cm²	40gm CO2	Powder Coated
2	Mechanical Foam	Foam AFFF	9 ltr.	90% min	6m min	25s-60s	30Kgf /cm²	60gm CO2	Powder Coated
3	Dry Chemical IS 4308	Dry Powder 10Kg	5 Kg	85% min	4m min	15s-20s	30Kgf/cm²	120gm CO2	Powder Coated
			do	6mmin	23s-30s	do	180gm CO2	do	
4	ABC Storage Pressure	Dry Powder (ABC type)	0.5 Kg	85%min	1.5min	5s-8s	30Kgf /cm²	-	Powder coated
			1.0 Kg	85%min	5 min	8s-10s	do	-	do
			2.0 Kg	85%min	2 min	8s-10s	do	-	do
			5.0 Kg	85%min	4 min	15s-20s	do	-	do
			10 Kg	85%min	6 min	23s-30s	do	-	do
5	CO2	CO2	2 Kg	-	-	8s-16s	250Kg /cm²	110Kgf/cm²	Not Reqd Mangnese
		do	3 Kg	-	-	8s-18s	do	85Kgf/cm	do
		do	4 Kg	-	-	10s-18s	do	do	do
		do	6 Kg	-	-	10s-20s	do	do	do
		do	9 kg	-	-	15s-36s	do	do	do
		do	22 Kg	-	-	20s-60s	do	do	do
6	Halotron 1	Helotron	0.5Kg	-	1.5m	5s-8s	-	15Kgf/ cm²	Not Reqd Mangnese
		do	1kg	-	2.0m	6s-10s	-	do	do
		do	5Kg	-	2.0m	8s-12s	-	do	do
		do	5Kg	-	3.0m	20s-25s	-	do	do
MOBILE									
7	Dry chemical	Dry powder IS 4308	25Kgs	85% min	6m min	25s-30s	30Kgf /cm²	1KgCO2	Powder coated
			50Kgs	do	8m min	40s-50s	do	2KgCO2	do
			75Kgs	do	10m min	50s-60s	do	do	do
8	Water CO2	Water	50ltr	95% min	10m min	60s-180s	30Kgf /cm²	300gm CO2	Powder coated
9	Mechanical foam	Foam AFFF	50ltr	90% min	10 m min	40s-180s	30Kgf /cm²	300gm CO2	Powder coated
MODULAR									
10	ABC Ceiling mount	ABC powder	5Kgs	-	-	-	30Kgf /cm²	-	Powder
		do	10Kgs	-	-	-			coated
11	Halotron 1	Halotron gas	5kg	-	-	-	30Kgf /cm²	-	Powder coated

Charged/ Expansion space pressure	Overall width	Overhall Height	Shell Diameter	Chargge Weight	Temperature Range	Operation Method	Type of fire	Material Creating Fire
11	12	13	14	15	16	17	18	19
-	290 mm Approx	570mm Approx	180 mm	14Kg Approx	27±5°C	Upright	A	Wood, plastic, paper, rubber, cloth
-	290mm Approx	570 Approx	180mm	14 Kg Approx	27±5°C	Upright	AB	Petrol, paints, other solvent
-	250mm Approx 280mm Approx	515 Approx 560 mm Approx	150mm	11 Kg Approx 19Kg Approx	27±5°C do	Upright do	BC do	Petrol, Paint (liquid), gases, Powder Elect.Equipment
15Kgf/ cm²	132mm Approx	245mm Approx	80mm	2.5Kg Approx	27±5°C	Upright	ABC	All Classes of fire
do	132mm	330mm	87mm	2.5kg	do	do	do	do
do	135mm	370mm	108mm	4.2Kg	do	do	do	do
do	185mm	510mm	150mm	9.6Kg	do	do	do	do
do	230mm	615mm	180mm	16Kg	do	do	do	do
-	-	550mm Approx	108mm Approx	7.6Kg Approx	27±5°C	Upright	BC	Petrol,Paint, liquid gas
-	-	540mm	140mm	13Kg	do	do	do	do
-	-	710mm	140mm	17Kg	do	do	do	do
-	-	940mm	140mm	22Kg	do	do	do	do
-	-	1225mm	140mm	38Kg	do	do	do	do
-	-	1150mm	232mm	66Kg	do	do	do	do
15Kgf/ cm²	-	245mm Approx	65mm Approx	1.45Kg Approx	27±5°C	Upright	ABC Type	All classes of fire
do	-	270mm	87mm	2.2Kg	do	do	do	do
do	-	320mm	87mm	3.3Kg	do	do	do	do
do	-	410mm	150mm	9Kg	do	do	do	do
15Kgf/ cm² max	325mm	1100mm	320mm	75Kg Approx	27±5°C	Upright	BC Type	Petrol,paint, liquid gas
do	440mm	1170mm	382mm	135Kg	do	do	do	do
do	540mm	1200mm	382mm	160Kg	do	do	do	do
-	-	1200mm Approx	320mm	100Kg Approx	27±5°C	Upright	A	Wood,Plastic, paper,rubber
-	1125mm Approx	-	320mm	100Kg Approx	53±2°C	Upright	AB	Petrol,paints, other solvent
15Kgf /cm²	-	350mm	230mm	9.5kgs	57/68/79°C	Ceiling mount	ABC type	All classes of fire
		do	300mm	16kgs	do	do		do
15Kgf /cm²	-	350mm	230mm	9.5Kgs	57/68/79°C	Ceiling mount	ABC type	All classes of fire

3.4 Fire Fighting System:

The system deals with extinguishing and preventing the fire which has occurred at some place. The obvious method for a fire to die out is to cut the supply of oxygen by surrounding fire with water or foam or non flammable non combustible gas. The fire at initial stage can be knock down by stream of water, forming thick blanket over the burning surface, chemical powder or injecting CO2 on the burning material. The equipment used for this purpose is called Fire Extinguisher, or fire fighting equipment.

3.4.1 Fire Extinguisher: The fire extinguisher can extinguish fire, if used at initial stage of fire, but it is ineffective after fire is broken out. So these are only first aid fire fighting equipment. There are three classes of material which gets fire, accordingly the fire extinguishers may be classified as A, AB, BC or ABC type.

A Type: They are best suited to combat fire causes due to burning of solid organic carbonaceous material like wood, paper, plastic, cloth and rubber.

AB Type: They are suitable for fire due to burning of voltaic flammable liquid like petrol, paints and other solvent.

BC Type: They are suitable for BC class of fire involving inflammable liquid like petrol, paints or gases.

ABC Type: These are used for all type of material and electrical appliance specially.

The fire extinguisher can also be classified according to the agents used for knocking out different class of fire.

(1) Water CO2

(2) Mechanical foam

(3) Dry chemical powder

(4) Store pressure dry chemical powder

(5) Halotron-1

The fire extinguisher can also be classified as portable, mobile or ceiling mount or modular type. The detail of specification and qualities of fire extinguisher are summarized in the following table. The bodies of these fire extinguisher are either deep drawn steel or seamless manganese. On deep drawn steel anti-rust substance coating is provided with proper post office red color painting.

The portable fire extinguisher are hanged on the appropriate wall of corridor, lobbies or staircase at suitable distance, so that one can easily and quickly asses the fire breaking point. These first aid extinguishers are used by occupants of building or fire officer to distinguish fire.

3.4.1.1 Estimating and Costing: For the estimation, the cost of supply and fixing of fire extinguisher can be found after proper design for type and location of fire extinguisher.

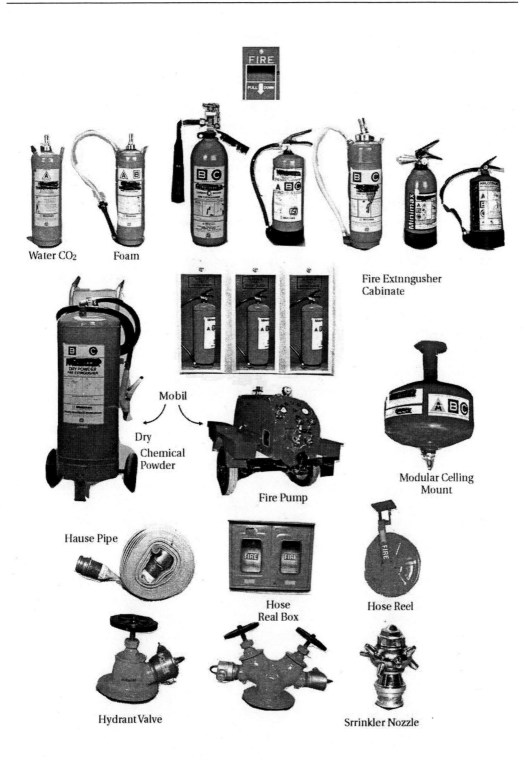

Water CO₂   Foam

Fire Extnngusher
Cabinate

Mobil

Dry
Chemical
Powder

Fire Pump

Modular Celling
Mount

Hause Pipe

Hose
Real Box

Hose Reel

Hydrant Valve

Srrinkler Nozzle

3.4.1.2 Design: The type of fire extinguisher is selected depending on the class of possible fire that may occur in the building or premises. The fire extinguishers are located in the corridor, lobbies, staircase etc. to be protected at suitable distance according to their construction. In general the distance between the extinguishers is approximately 15m. Thus selecting the location and type of fire extinguisher the number of each type can be found and summarized in the schedule of material.

3.4.1.3 Detail of Measurement: The table called schedule of material or item is prepared consisting the different class of extinguisher with their location. Their total will give the quantities of each class of extinguisher.

3.4.1.4 Analysis of Rates: Analysis of rates of each class of extinguisher are prepared on their current market rates as per price list of manufacturer, if their exist no schedule of rates.

3.4.1.5 Bill of Quantity: Knowing the quantities and rates of each extinguisher the bill of quantity is prepared on the standard format calculating the amount of each item. The total of the table will give the cost of supply and fixing of fire extinguishers.

3.4.2 Hydrant or Sprinklers Fire Fighting System: The fire is made to extinguisher by using jet of water or sprinkler, using fixed or sprinkler nozzles. Nozzle for fixed jet or sprinkler jet requires hugs amount of pressure of water which is created by pumps. These system requires huge amount of water so we construct and maintain storage water tank. It may be underground or overhead at the top of building. The overhead tank is preferred when it is not possible to provide water pump in multistoried building. In this case the pressure for hydrant at top floor is very poor to combat fire, though overhead tanks are useful for water sprinkler at basement of building.

3.4.3 Selection of Two Fire Fighting System: Various building bye-laws, code of practice, fire act prevailing in the area, in which building is situated, have given guidelines to select the type of fire fighting system. As per Delhi bye laws 1983 or revised, prevalent in near by state also, following recommendation for selection of these two system has been made mandatory.

Sl. No.	Type of building	Type fire fighting system
1.	Residential building of any floor area less than 15m in height.	Nil
2.	Residential building of more than 15m height	Hydrant system
3.	Non residential building of height 15m or above and of any floor area.	Hydrant system
4.	A basement or sub basement of building of 15m or more height when hazardous or explosive materials are stored.	High pressure water or foam sprinkler

5.　　All building of above category when risk of fire or　Automatic water or
　　　loss is more　　　　　　　　　　　　　　　　　　　foam sprinkler

Depending on the conditions both type of fire fighting system can be used i.e. sprinkler for basement and hydrant for other part of building.

3.4.4 Hydrant Valve System: In this system the fire is extinguished by jet of water by using hydrant valve and nozzle. The water at pressure by water pump or gravitational pressure due to overhead tank, is supplied through GI or MS C class pipe of proper diameter (4", 5" or 6"). The diameter of pipes is designed as per height of building, so that water pressure at hydrant valve is maintained at required pressure, with out increasing the pressure in risers exceeding the limit to damage the pipe. The diameter of pipe (T-pipe) used to connect hydrant is 100mm. The different equipment accessories used with the system are as follows.

3.4.4.1 Wet or Dry Riser: A vertical GI or MS pipe used in building to supply water from water pump to hydrant valve at different floor is called riser. The water is kept at pressure in the riser. There are two type of risers. One, in which water is always available at required pressure are called wet risers. The other pipes without water in normal condition, but water is supplied at required pressure by fire pump, at the time, when fire is broken and when it is required to combat the fire, are called dry risers. Pressure in riser is so maintained that water pressure on highest hydrant valve must be not less than 1.5 Kg / cm² and first aid hose reel may discharge 1140 liter / minute at 1.5 Kg / cm² pressure. In building of height from 15m to 60m, 100mm GI or MS pipe is used to feed upper hydrant valve. If building is above 60m another riser is used as per following conditions to avoid excess pressure.

Sl. No.	Riser	Height of riser / building	diameter of riser
1.	1st riser	Up to 60m	100mm
2.	2nd	60m to 100m	125 mm
3.	3rd	100m to 150m	150mm

3.4.4.2 Down Comer: If fire pump is not used, the vertical pipes, in which the water is supplied from overhead tank because of gravitational force through a gate valve to hydrant valve at the time of fire, are called down comer.

3.4.4.3 T Connection Pipe: The pipe connected to riser or down comer and hydrant valve and in shape of T, is called T connection pipe.

3.4.4.4 Hydrant Valve: The special designed outlet valve, which has two valve to stop or open the flow of water to the hose pipe with nozzle to form water jet to diminish fire is called hydrant valve. There are two type of hydrant valve.

(a) Internal Hydrant valve:- The hydrant valves used inside building at every floor for combating the fire inside of building at that floor are internal hydrant valves.

(b) External hydrant valve:- The hydrant valves used out side building to combat fire from outside building are called external values. These are also called courtyard fire hydrant valve.

The hydrant valve is kept in a glass panel so that it may be clearly visible and in case of non availability of key of glass panel it may be broken to use this hydrant valve. A hose reel on swinging frame is also provided in these glass panel. A hydrant valve is twin type and with two opening fitted with female coupling. One opening is connected with hose pipe with adopter coupling and other opening is used to connect hose reel.

3.4.4.5 Non Return Valve: A non return valve is provided at the outlet of fire water pump or in suction pipe at the outlet of sump or underground water tank and at the bottom of the riser. It is also provided at outlet of the fire brigade connection at water tank or riser and the outlet of terrace storage tank or fire pump so that water may not return back, if fire pump is put off, to maintain pressure in pipe line i.e. ring or risers.

3.4.4.6 Pressure Reducing Valve: A pressure reducing disc valve is provided at the outlet to the external hydrant valve so water pressure may not increase more than 5.5 Kg / cm².

3.4.4.7 Air Valve: Except in residential apartment buildings, every rise is extended upto terrace and an air valve is provided to remove air, if air or air bubble is trapped in riser.

3.4.4.8 Gate Valve: Gate or stop valve is provided at the outlet of fire pump and at any other required places.

3.4.4.9 Painting: Wet or dry risers, down comers and all water pipes hydrant valves, T connection pipes, pipe ring or feed pipe, hydrant boxes etc. are painted with post office red color paint.

3.4.4.10 Hose Reel & Hose Reel Box: A 19mm, 30m swing type rubber hose reel on pulley wheel mounted on frame is kept in hydrant glass panel box or separate hose reel box with glass panel. Hose reel box frame & pulley etc. are also painted with post office red color paint.

3.4.4.11 Fire Hose: A 100mm rubber hose pipe 30m long or two 15m long rubber hose pipes with joint coupling arrangement is also kept in hose box with glass panel to reach whole area on every floor from hydrant valve to combat fire. Accordingly the number of risers are selected.

3.4.4.12 Fire Pumps: The water to riser is supplied with fire pump. This is a diesel engine driven multistage centrifugal pump. If there is D.G. set on the site of building or in fire pump house, an electrical motor pump set is provided, for better and effective fire fighting operation. Both type of pumps may be provided, so that any of them can be used in case of other's failure. The discharge of these pumps must be 2275 liter per minute. When number of hydrant valves are opened or used, the pressure and discharge of water in riser at hydrant value falls, a pressure switch fitted in water pipe line actuate booster pump called jockey pump through relay and contactors on fire fighting panel. As pressure in riser is maintained, the jockey pump automatically put off. The jockey pump is also a multistage centrifugal pump. The HP of the fire pump is calculated from water discharge as below. The horse power of either diesel engine or electric motor pump is taken as same as calculated but jockey pump is taken of less HP according to number of hydrant valves and range of falling pressure in riser.

$$
\begin{aligned}
\text{Hose Power (HP)} \quad &= \quad W \times Q \times H \,/\, 75 \times E \\
\text{Where} \quad W \quad &= \quad \text{Specific weight of water } (1000 \text{ Kg} / \text{m}^3) \\
Q \quad &= \quad \text{Discharge of water in m}^3 / \text{sec.} \\
H \quad &= \quad \text{Total head or water in meter} \\
E \quad &= \quad \text{Over all efficiency}
\end{aligned}
$$

Normally discharge Q is taken as 2275 liter per minute in fire fighting system, H takes losses (friction etc.) into account, so H is slightly more than the height of building, E is taken as 48 to 50%.

So for a building of height 50m

$$
\begin{aligned}
H \quad &= \quad 60\text{m} \\
E \quad &= \quad 50\% \\
S \quad &= \quad 2275 \text{ liter} / \text{minute} \\
&= \quad 2275 \times 10 \,/\, 60 \text{m}^3 / \text{s} \\
&= \quad 37.9 \text{ m}^3 / \text{s} \\
\text{So HP} \quad &= \quad 1000 \times 37.9 \times 10 \times 60 \,/\, 75 \times 0.5 \\
&= \quad 60 \text{ HP}
\end{aligned}
$$

3.4.4.13 Water Supply: As discussed earlier the water is supplied to risers by fire service pump which takes water from under ground storage tank. The water tanks are filled either by main water works line or from own tube well. The required water rate is 1000 liter /min. at the time of continuous use of water tank when fire fighting operation is carried out. Accordingly the water is provided with proper size of manhole for filling of water with hose & mobile water tank, service, cleaning, water treatment or other maintenance purpose. The slab of under ground water tank is made to bear 18 ton load of mobile water tank etc. The capacity of the sump is decided as per height & occupancy of building in following manner.

Sl. No.	Type of the building	Sump capacity
1.	Residential building apartment of height 15m to 24m	50,000 liter
2.	Residential building apartment of height above 24m	1,00,000 liter
3.	Non residential height 15m to 24m with average occupancy e.g. office, shopping complex or departmental store.	1,00,000 liter
4.	Non residential building of height 24m or more	2,00,000 liter

3.4.4.14 Terrace Water Storage: For fire fighting purpose in addition to sump, water storage is provided at terrace of the building in a terrace water tank of 20,000 liter capacity. It may be of more capacity as per actual requirement on advice of fire officers. Wet risers are also connected to these terrace water tank through a non return value so that in case of emergency it can be used either in beginning or on failure of fire pump service.

3.4.4.15 Modification in Storage Tank: As the underground water tank sump is used only in case of fire fighting but water tank is kept full for use at any time fire Water in storage tank weather under ground or terrace remains stagnated for long time to give bad smell. So it is preferred to construct modified storage tank i.e. tank for general water supply is construed in two compartment by making a partition valve of suitable height so the capacity of storage for fire fighting become as per minimum requirement. Generally water supply portion is filled by over flow water from fire fighting portion as shown in figure. The pipes connection is accordingly made.

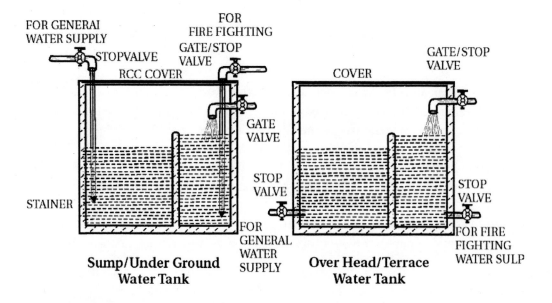

Sump/Under Ground Water Tank

Over Head/Terrace Water Tank

3.4.4.16 Fire Brigade Connection: In case of emergency on the failure of fire service pumps or water tanks being emptied out (or no water is left out in tank), the water is supplied by fire brigade tank or water is required by fire brigade from water tanks at building so fire brigade connection are made at sump or in riser as described following.

(a) Fire brigade inlet Siamese connection is made near water storage tank at ground to supply/ receive water by fire brigade tanks. It is 150mm GI pipe with four flanged hydrant valve, connected to storage tank. The water is sent/received through fire brigade tankers hose connecting to hydrant valves.

(b) Other fire brigade connection is made in wet riser system, so that fire brigade may directly supply water automatically from their fire service pumps to the wet riser.

3.4.4.17 Operation of Hydrant System: On being located of fire by detection system the fire staff after evacuation building by occupant, start combating operation by connecting hose pipe with nozzle with hydrant valve by adopter coupling (male) with female coupling of hydrant value. As per requirement the hose real may also be used with other opening of hydrant valve. As soon as hydrant valve is opened and water is being flown, the pressure inside wet rise falls, which actuate pressure switch in riser system actuating current / signal, start fire pump through relay and contactor at fire panel automatically. The operation of fire pump may be manually started on listening call from fire staff at hydrant valve. A hooter is also provided with fire fighting system, which indicates automatic start of fire pump. The switching off of pump motor is always done manually after completion of operation.

If many hydrant valve are used to use number of hose pipes with nozzle, the pressure in riser falls, the required value, which starts standby jockey pump through another pressure relay & required contactor in addition to main pump. This pump automatically put off as pressure reached required value in riser.

3.4.5 Automatic Sprinkler System: This system is provided in basement or sub basement where car parking, store of flammable material, laundry etc. are situated or on the floor where departmental store, shop or trading is done i.e. fire risk is more. This system is not needed in a building provided with smoke heat detector and where risk is not very high and fire trained staff is available 24 hours.

3.4.5.1 Pipe Line: GI or MS pipe 25mm to 100mm C class are used for sprinkler circuit depending on the number & place of fire sprinkler such that pressure at lowest sprinkler is 1.5 Kg / cm². Sprinklers are connected to water pipe through sprinkler head, which are filled with pressurized water. The water at pressure is supplied by sprinkler pump and underground sump and terrace or over head tank only. NRV and gate valve are provided as per requirement similar to PRV, air valve at storage tank.

3.4.5.2 Sprinkler Pump: This is an electric motor driven by horizontal centrifugal pump (same as in hydrant system) with discharge 625 gallon / min. and total head 230 ft with suction side velocity 300ft / min. Its suction side discharge can be found by multiplying it with area cross section of pipe.

Both hydrant pump and sprinkler pumps are interconnected so they can be used as stand-by of each other. Both are placed in same pump room. These pumps are provided in basement of building so that self priming of pump is done. If basement is not provided in the building or it is not possible to make pump room in basement, the underground pump room is made for self priming of pump.

3.4.5.3 Operation of Sprinkler System: Sprinkler head has a glass bulb filled with chemical liquid having much coefficient of expansion. On broken of fire, the temperature of surrounding increases expanding liquid chemical, so the glass bulb is broken at fixed temperature (70°C in general) opening the orifice of water pipe connected to sprinkler having water at pressure, which start water spray of sprinkler automatically subsiding fire immediately and saving huge damages.

3.4.6. Estimating & Costing: To prepare the estimate & to find the cost of supply and installation of fire fighting system, we design the fire fighting systems and their various components as described as follows.

3.4.6.1 Design of Fire Fighting System: As per type, importance, occupancy & height of the building and its risk to fire, the type of fire fighting system is selected as discussed earlier. In both system water storage tank, its capacity, the size (HP) of fire pump and size of pipe are worked out.

3.4.6.2 Design of Hydrant System: The number of riser and hydrant valve connected to riser can be worked out knowing floor area, length of corridor or passage, so that staff can reach any place with 30m hose from each hydrant valve. A hydrant valve is connected to each riser at every floor via T connection pipe. In general one riser is required per 930sqm floor area. For building of height up to 60meter 100mm dia pipe riser is required but for building of height greater than 60m and less than 100m, 125 dia pipe riser is required which is separate from ground and is 2nd riser per 39sqm floor area. Similarly building of height 100m to 150m 3rd riser of 150mm dia pipe is required e.g. Twin tower (Kolalumpur) which has 88 stories and its height must be more than 250m multi-risers of suitable dia pipes are used in such building over head terrace tank is used with under ground tank.

Once number of hydrant valve worked out than the number and size of the riser can be found. No horizontal pipe is allowed for water to flow up to hydrant. As hydrant valve is connected through T connection pipe. Thus number of T connection pipe, hydrant valve and length of MS / GI pipe of different diameters can be calculated to give detail of measurement. The numbers of gate valves NRV, PRV, air valve are counted. Knowing hydrant valve the number of hose real, hose real box and 30m hose pipe are found. Fire brigade connection (siemese), one for a storage tank and other type of connection with gate valve and one NRV per riser is provided.

3.4.6.2 Design of Sprinkler System: Fore every 10sqm (100sft) floor area, one sprinkler is required. So number of sprinklers can be found considering discharge & pressure at sprinkler. Size (diameter) of pipe is calculated, then the length of pipe for down comer / riser, pipe circuit from pump hose to sprinkler is found for each diameter pipe. The number of gate valves, NRV are also worked out.

3.4.6.3 Detail of Measurement: On the basis of above design the detail of measurement showing place and different components is prepared in a tabular form and totaled to give quantities of each item.

3.4.6.4 Analysis of Rates: On the basis of manufacturer price list or current market rates the analysis of rates of each item is prepared, if schedule of rates of these items are not available.

3.4.6.5 Bill of Quantity: The bill of quantity to find the cost for providing Fire fighting system is prepared on standard format, if the quantities, rates of each items (equipment, device, component or accessories of the system) are known. The total of bill of quantity will give total cost required for installation of fire fighting system

3.5 Summary of Cost: Knowing the cost of works, as per their bill of quantities, for Detection and fire alarm system, Fire fighting system, Fire extinguisher and PA system, if required, the summary of cost or abstract of cost is prepared to find complete cost of the Fire detection and protection system. Contingencies charges for miscellaneous expenditure, centage charges if required, are also added to get the cost of estimate or DPR.

4 | Sound System

4.0 Introduction:

The sound is a form of energy. When a human being speeks a sound is produced. The sense of speech is converted into sound by the mouth including nose and neck and travels in the form of sound wave. The sound wave travels in any matter solid liquid or gas as transverse or longitudinal wave with the help of vibration of particle of the medium of travel. The sound wave can be reflected or refracted when traveling from one medium to other. Reflection and refraction depends on density of medium. Reflection dominates over refraction as sound wave enters in denser medium from rare medium. The sound is reflected by any object at distance from sound source as in a valley by solid hills, to give echo. Sound may be pleasant or unpleasant hearing e.g. musical sound is pleasant but noise is unpleasant. The sound wave when strikes with diaphragm of ears of human being, it gives sense of listening. This is how a conversation of human being takes place.

Sound is associated with vibration. It is produced, travels or transmitted and reproduced with the helps of vibration. As sound travels in all direction, so it decays in traveling distances. Sound of natural voice of a human being is weak to travel and to be audible at long distances, some system of produces of boosting sound is required. Such sound system is called loud speaking system. Sound system needed to be installed in different building is of greater interest for any engineer associated with construction of buildings, especially with electrical installations in buildings.

4.1 Terminology and Defination: It is useful to converse with some of the following terminology before proceeding further.

4.1.1 Pitch: Pitch of sound is related to frequency of sound i.e. greater the frequency, higher the pitch of sound (Heavy or Thin voice of human being)

4.1.2 Naturalness: Naturalness of reproduced sound means that it should not be harsh or boom or have its tone altered but as good as natural voice or sound. There are three factor affecting naturalness and faithfulness.

1. Acoustics of space in which sound source is placed.
2. Characteristic of amplification equipment i.e. microphone, amplifier and loudspeaker.
3. Acoustics of space where sound is reproduced.

4.1.3 Directional Realism: A listener should perceive that the amplified sound is coming from the actual location of source if a performance is carried out on a stage, in front and reinforced sound is coming from loudspeaker in rear, then directional realism will be missing.

4.1.4 Intelligibility: The listener must clearly and easily understood the sound coming to him, otherwise the purpose of sound system installation is meaningless e.g. announcement at railway station must be of sufficient intelligibility to be clearly heard by passenger on station.

4.1.5 Loudness & Intensity: A sound of large amplitude shall have large loudness and grater intensity, if L & I is loudness and intensity then

$L \alpha \log I > L = K \log I$

Intensity level of sound with reference to 0 level L o, is defined as

$L = L_1 - L_0 = K (\log I_1 - \log I_0)$

$$= \log I_1/I_0 \quad \text{taking } K = 1$$

Where L_0 is the intensity of 0 level & L_1 is intensity of sound at any place.

If natural logarithm is used i.e. at base 10 and if $I_1 = 1 U I_0$

So $\quad L = \log I_1/I_0 = \log_{10} 10 = 1$ bel

Bel is the unit of sound loudness named for Alex Grahm Bel inventor of telephone. But normally sound loudness is measured in decibel (dB).

$$1 \text{ decibel} = 1/10 - 0.1 \text{ bel}$$

The loudness of sound is expressed as sound pressure level SPL & is measured in dB. The following table gives the loudness of sound and its effect on human being.

SOUND	LOUDNESS (dB)	EFFECT
-	0dB	-
Normal breathe	10dB	Barely audible
Rusting of leaf	20dB	Audible
Whisper	30dB	Very quiet
Mosquito buzzing	40dB	Quiet
Quitet home	50dB	Quiet
Normal conversation	60dB	Loud
Busy traffic	70dB	Loud
Heavy truck	90dB	Loud
Thunder	100dB	Loud
Siren	120dB	Painful threshold
Machine gun	140dB	Painful
Jet take off	150dB	Painful
Large rocket	180dB	Painful

The temperature and heat affect the loudness of sound (Binural effect).

4.1.6 Ambient Noise Level (ANL): Noise is present every where in universe due to presence of many source of sound, the loudness of the resultant of these sound is measured and termed as ambient noise level of area.

4.1.7 Requisite Loudness : It implies that the amplified sound be heard comfortably and with clarity over the ANL present. The requisite sound pressure level must be at least 10dB more than the ANL level at all point in the listening area. Though there is limit to sound pressure level which can be achieved by amplifiers, if microphones and loud speakers are placed in same acoustic environment, due to feed back or howling. This can be minimized by acoustic treatment and controlling volume of amplifier.

4.1.8 Echo & Reverberation: A sound received later after reflection of sound from its source after sowe delay as compared to the direct sound wave hearing. This is called Echo. When direct and reflected sound superimposes each other maxima or minima zone of sound is produced. This is called reverberation and is evidenced by persistence of sound even after the source of the sound is ceased.

If V is volume of the hall, a is coefficient of absorption of surface of area S then for good acoustic.

$$\text{Reverberation time} \qquad T = \frac{-\ 0.05V}{\Sigma a \times S}$$

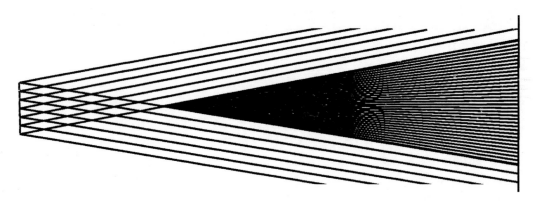

Fig. 4.1.8

4.1.9 Acoustic: It is a branch of science which deals with production, reception and propagation of sound. There are three type of acoustic.

(i) Acoustic of building construction or architectural acoustic.

(ii) Electro acoustic - It deals with microphone, amplifier and loudspeaker.

(iii) Musical acoustic - It deals with musical sound.

4.2 Acoustic of Building:

Hall or any big room of building must be designed so that a person wherever he sit in the hall, he must have the privilege of hearing speaker effectively i.e. intensity level at any place or corners must be satisfactorily. It can be achieved by making a stage wall parabolic, avoiding any curved construction of the wall of hall. The factors affecting acoustic are,

(1) **Reverberation Time:** It must be optimum e.g. 0.5 seconds for speech and 1.0 to 1.5 for truck

(2) **Loudness:** Proper loudness even at the rear hall is maintained by placing good reflectors at proper place. Ceiling is made curved to increase reflection and reduce absorption.

(3) **Undue Focusing:** Undue focusing must be avoided by increasing open window and good absorption.

Seating man is equivalent to 0.5 m² of an open window.

(4) **External Noise:** It is received by open window or door structure. Spongy material is used on wall or ceiling to absorb external noise.

(5) **Placement of Window, Reflector or Absorber:** So after constant acoustic study is made, the reflectors or absorbers are placed as below :-

(i) To increase absorption seats must be more as seating man is equivalent to 0.5 m² of open window.

(ii) Rough cloth curtain are placed when reflection/projection are in wall construction.

(iii) Painting or maps in thick glass is pleased on wall.

(iv)Ceiling or wall is made rough to absorb and covered with porous material like perforated thermocol thick felt and canvas etc.

(v) Seat with cushion is provided and must be properly upholstered as when man is sitting

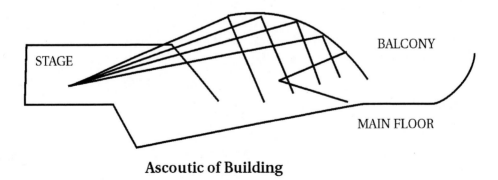

Ascoutic of Building

on seat it, act as absorber and when man is not sitting, it act as reflector.

FIG. 4.2

4.3 Decay:

As sound travels due to expanding of its wave front the intensity or loudness of sound decays. In free field area without sound reflecting surface, sound intensity decays according to distance it travels. The decay or attenuation in sound intensity or sound pressure level SPL is 6dB for every doubling of distance i.e. If sound level is X at 1 m. from source, it will be (X-6) at 4m distance.

4.4 Power Change Effect:

When power of loudspeaker is doubled, the incrase in SPL is going to be 3 dB and when increase in power is 10 time, the increase in SPL is 10dB. Also if loudspeaker of same sensitivity are clubbed together, the increase in SPL is 3dB when the quantity of speakers are doubled.

4.5 Loud Speaking System:

This system can broadly be divided into two system, namely, Public Address System and Reinforcement System, beside some more different kind of sound system.

4.5.1 Public Address System: In this system sound surce and the listener are at distance, may or may not be in same acoustic environment. Its main requirement are intelligibility of speech and sound to be loud enough to overcome high ambient noise level (ANL) e.g. Airport, Railway station, Bus terminal, Sports stadium, Shopping malls (uses both public address & musical system) hotel (car calling), industrial complex & factories, exhibition ground public meting and rallies and mobile P.A. etc.

4.5.2 Reinfocement System: In this system sound is amplified or reinforced in the same acoustic environment as the listener. Naturalness, directional realism requisite loudness are main requirement of this system. Its example are theatres, auditorium, lecture hall, places of worship, conference venue, coffee house, restaurant, bar etc. Other sound system, which may be beside being either public address or reinforcement system used for some purpose too are enumerated below.

4.6 Conference System:

This system is used to reinforce sound of the personal / delegates sitting in a meeting or conference, which can not be heard, if they conversant or confer without conference sound system. Beside reinforcement of sound it can be used to minutes or record the proceeding and to play back recorded sound.

4.7 Communication System:

If the person are sitting at long distances, may be in same building or not, this system is used to reinforce, transmit sound through wire (Telephone) network or without wire (Cordless or Mobile Telephone). This system besides using basic equipment of loud peaking system, uses modulator, oscillator, microwave as carrier, transmitter receiver and telephone exchange.

4.8 Recording:

Instead of direct communication or direct listening we can listen a sound after some interval or a laps of a time. For this the sound is recorded and reproduced later. Audio tape records, CD cassette, DVD are the different version of recorded sound which is produced by a sound recorder and reproduced by a sound player e.g. tape recording, CD and DVD writer or player etc. Video recording is also done beside audio recording. Recording may be one of the following kinds.

1. **Disc or Mechanical:** The sound is produced before a microphone which creates electrical pulses of same frequency that of sound. These pulses make a needle to vibrate to cut wave pattern on a record. By electroplating or thermo pressing we can produce may record (Negative copy) from master. By a gramophone the process is reversed to reproduce sound using speaker in place of microphone.

2. **Optical or Film Recording:** Sound vibration after amplified by a amplifier via microphone as a electrical wave is put to a especially made bulb which blows variably as per intensity of sound coming to it as electrical pulse. The fluctuation of light is focused on a photo graphic film to form bright and light area in the film. The process is reversed for reproduction of sound using speaker in place of microphone.

3. **Magnetic Recording (TAPE):** It is conventional and common now-a-days. The sound spoken before a microphone is converted into electrical current of verying intensity which passes through a coil of a electromagnet causes varying magnetic field. Cellophane tape covered with magnetic oxide or iron is used to record the magnetic variation as the variation of atom in oxide on tape, which is moved at a constant speed by a electric motor against the magnetic head (An special electromagnet). The process is repeated for reproduction of the sound by using speaker in place of microphone.

These processes of recording can be shown in a line diagram as below.

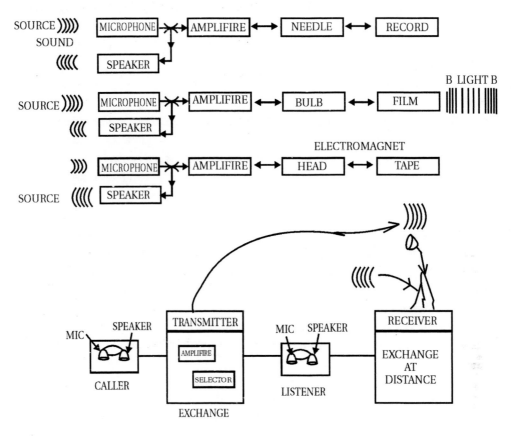

Communication System

FIG. 4.7

4.9 Broadcasting:

If a person want to listen the sound of live program or speech of a long distant person, the sound is broadcasted in atmosphere by a transmitter and is reproduced by a receiver such as radio or transistor. This sound system is used in association with recording, and reinforcement system is known as broadcasting sound system. It may be video or audio e.g. television or radio transmitter station.

4.10 Film or Video Production:

Sound system plays vital roll in film or video clips production together with camera as it governs the sound effect, dialogue, musical performance. This system consists of reinforcement public address systems with recording system both audio and video is termed as film or video production system. This system uses at television and film studios.

4.11 Musical Performance:

In a musical performance system, an artist whether vocal or instrument player is supplemented with sound system. Vocal song or sound produced by an instrument is audible up to a certain distance only and the reinforcement system is needed for a listener at distance to hear the music with quality and clarity, such system is known as musical performance sound system.

4.12. Basic Equipments:

Before the detailed description of various sound system, we shall discuss basic equipments needed in all loudspeaker system or simply sound system.

The basic equipments used in loud speaking system are (1) Microphone (2) Amplifier (3) Loud speaker.

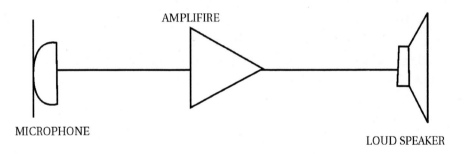

FIG. 4.12

4.12.1 Microphone: As such it is difficult to increase sound energy i.e. sound level to travel long distance range, so it is converted into electrical energy in the form of electrical pulse or signal. The electrical energy can be amplified or signal can be modified easily as compared to sound or acoustic energy. The device, which convert acoustic energy into electrical energy is called microphone. It converts our sound into equivalent electrical signal, though it is too feeble to drive a speaker.

Microphone are of different kind depending upon their application i.e. used for vocal or instrument, public address, conference or lecture, used with cord or cordless, with table, desk or floor stand or without stand. Every microphone has a cartridge with a on-off switch in a body. The cartridge consists of a diaphragm, magnet & a coil. Sound waves strike the diaphragm causing vibration in air core of magnet resulting change in magnetic field of the magnet resulting in current or electrical pulse or signal in the coil present in the magnetic field. Microphone can broadly be classified two types i.e. dynamic or condenser microphone.

4.12.1.1 Dynamic Microphone: There are simplest and rugged microphone and do not require external voltage source. These can withstand very high (loud) incident sound without distortion.

4.12.1.2 Condenser Microphone: These microphones require external voltage source for operation. They give higher output as compared to dynamic microphone with flat frequency response. Because of their component they can be built small and even miniature microphone using a condenser cartridge.

These two type of microphone can be further classified according to sound pickup pattern (polar response) or directional characteristic into three classes. (1) Omni directional (2) Cardioids (3) Super cardioids.

(1) Omni Directional Microphone: These pickup sound from all direction i.e. 360 coverage pickup. These small tie clip microphones near to mouth (on collar) are suited for hand free operation with ease of mobility of user.

(2) Cardioid Microphone: The coverage angle is 135° and these pickup sound from front, rejecting sound arriving from the back. Thus reducing howling and allows more volume opening of amplifier. These G.M. series microphone are suitable for podium mounting, conferencing, distance miking as well as miking for instruments.

(3) Super Cardioid: These pickup sound at a narrow angle in front with coverage angle 115° so these can be more effective in reducing howling and feed sound i.e. from audience and reflection from wall etc. So we select microphone as per requirement and decide the specification i.e. directional characteristic, frequency response, impedance sensitivity. The microphone frequency ranges are selected to match with frequency of sound source. These (microphones) transducer have low or high output impedance to match with amplifier. Microphone are also selected according to sensitivity mV/pa. In general we use microphones of 1.3, 1.4 or 1.5 sensitivity but higher sensitivity are used for professional artist.

4.12.2 Amplifier: This device uses to amplify the feeble electrical signal of the microphones sufficient to drive speaker for reproduction of original sound. This consist of a electronic amplifier PCB circuit board, in a housing (Body) Two transformer at input & output of circuit are required for power supply and impedance matching with speaker.

An amplifier may be single zone or multi zone output to feed greater number of loud speakers at different location to meet any type of public address sound system or reinforcement system requirement. An amplifier for general purpose or professional purposes can be classified as per out putt power, zone output, input channel and other specification of amplifier. Amplifier zone are worked out for bigger area i.e. left and right side, front, center or rear side of a hall/ground number of input channel of microphone and accessories. Some amplifier has built-in arrangement for CD player and AM/FM tuner.

4.12.2.1 Mixer: For reinforcement system different type of microphones are fed to a amplifier for mixing the different signal of microphone to give undistorted unit signal to the amplifier, this unit is called mixer.

4.12.2.2 Booster Amplifier: The gain of signal decays if the length of cord or wire used to feed distant speakers, is long the gain of amplifier is further amplified by a amplifier called Booster.

4.12.2.3 Equalizer Amplifier: It is an amplifier which can amplify particular frequency, if the output consist of sound of many frequencies after mixes and amplifier unit e.g. if we are listening music having vocal song together with the sound of many instrument we can reduce or increase gain of one or two instrument keeping the gain of other sound constant. This is very useful for musical performance sound systems to give more appealing or pleasant music or sound.

4.12.3 Speaker: Speaker is a transducer or device to convert electrical energy in to acoustic or sound energy and use to reproduce sound received by microphone which sends the signal to the speaker through an amplifier etc. It also consists of magnet, coil and diaphragm like that of a microphone, but bigger in size and its function/operation is reverse to that of a microphone. Amplified electrical pulses pass through the coil in a magnetic field, which changes flux in air core of the coil causing vibration in air column of core by which diaphragm vibrate to produce original sound.

There are two type of speaker i.e. cone type or compression type.

4.12.3.1 Cone Type Speaker : These type of speakers reproduce high quality sound with a wide frequency response and low distortion but it has low efficiency i.e. it convert same electrical energy in to low sound.

4.12.3.2 Compression Type Speaker : These speaker reproduce very high level sound but with limited frequency response. These speakers are used along with horn like WFA/WFS etc. to enable compression speaker to operate at high efficiency. A special compression type speaker is HF compression drive to extend the frequency response up to upper limit of the audible frequency range.

4.12.4 Lou6d Speaker: Both cone or compression type speaker can not be used alone, the cone type speaker is mounted on a baffle or cabinet and compression type speaker are coupled to a horn to get desired response or sound. We term it as loud speaker. So a speaker with cabinet or horn is called loud speaker. There are following type of loud speakers commonly used. (1) Horn Loud Speaker (2) Wall Mounted Loud Speaker (3) Column Loud Speaker (4) Ceiling Mount Loud Speaker (5) Box Loud Speaker (6) Subwoofer.

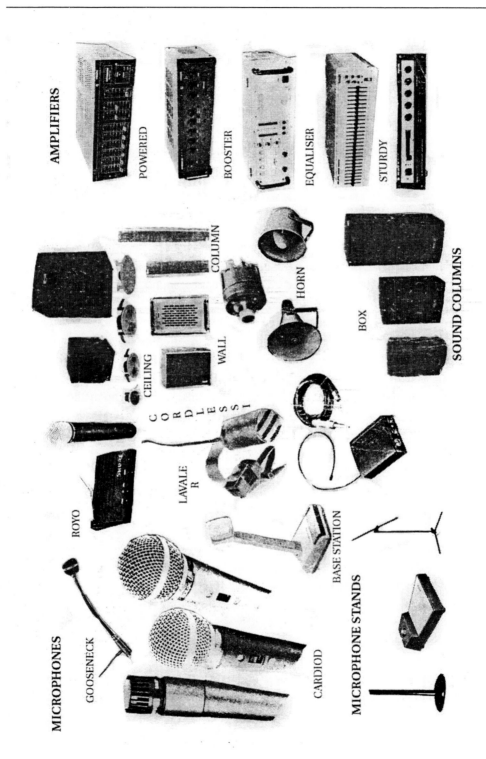

4.12.4.1 Horn Loud Speaker: These are commonly used since quite much earlier days especially out door. This loudspeaker consists of a compression driver unit coupled with suitable PA horn. These have high conversion efficiency so these are capable of producing much loud sound for the same input power as compared to cone driver's loud speaker.

Because of being limited frequency response they are mainly used in PA system. These can also be installed indoor where ANL is very high, such as in power generating station or industries with heavy machine noise. Horn can be exponential type (Trumpet) or reflex type (Re-entrant). Reflex type is smaller in size than trumpet type in design. Larger the diameter of the horn, batter is low frequency response.

4.12.4.2 Wall Mount Loud Speaker: These are compact full range or two way systems and is ideal for music, paging etc. and at places where there is no ceiling, no false ceiling or has high ceiling. So ceiling mounted speaker produce the desired SPL at the listening area.

The dispersion of such wall mount loud speaker is conical with 90° dispersion angle and work best when placed on facing walls and staggered so that each speaker is aimed at a pint midway between speaker. The idle spacing between two wall mount speakers depends on the height at which they are mounted and distance of facing walls where speaker are to mounted.

4.12.4.3 Column Loud Speaker: They consists of a number of cone drivers mounted one above other on a single baffle so the sound from a column speaker is projected forward in a flat beam as sound of each cone reinforce each other and not in the shape of a cone as in wall mount speaker so the intensity of sound and its range are greater in column loud speaker than conventional single loud speaker of same power rating. Maximum area of reinforcement occur at increasing distance from the column and this counter act, the fall of SPL with distance. Column speakers are used in poor acoustic condition. The outstanding features are (i) Intelligible reproduction (ii) Parabolic wide distribution (iii) Long throw coverage (iv) Compact and stern (v) Weatherproof (vi) Selectable wattage tops.

4.12.4.4 Ceiling Mount Loudspeaker: They provide uniform coverage and satisfactory frequency response for live speech environment as wall as background music application. The sound dispersion pattern is also conical being cone speaker, with 90° dispersion angel. As the ceiling height increases, coverage area of ceiling speaker increases, so power required by speaker to give same sound level at listener position also increases. For ceiling above 25 feet column or wall mount speaker are used instead of ceiling mount, at suitable height. In a good installation placement and spacing of loud speaker should be such that the variation in SPL through out the targeted listening with in 3dB.

Thumb rule for spacing of ceiling mount loud speaker depending on different application is giving blow.

Application	Spacing Between Adjecent Loud Speakers
Speech	0.6 × ceiling height
Background Music	1.2 × ceiling height
Speech & Music	1.0 × ceiling height

4.12.4.5 Box Loud Speaker: They are versatile and capable of reproducing high fidelity sound at very high SPL and used for touring sound. These loud speakers are one way, two way, three way system. The most common is two way system, which consist of low frequency cone drive and high frequency compression driver coupled with horn of constant directivity. Trapezoidal shaped cabinet are used as they easily stacked to give uniform coverage. These are used for sound reinforcement system indoor as wall as outdoor application in place of horn loud speakers.

4.12.4.6 Subwoofer: Subwoofer are mainly used, where even the lower audible notes are to be reproduced with full powers subwoofer loud speaker system. Subwoofer are used for reproduction of super bass range (Low frequency blow 200 Hz). They are combined with loud speaker system described earlier to enhance the bass response of the overall system e.g. music reproduction.

4.13 Public Address System:

As discussed earlier the system is used at, Airport, Railway station, Bus terminal, Seaport, Shopping Malls, Sport stadium, to address gathering by a speaker, at college or school, factory & industries for paging announcement, for car calling or any announcement in a hotel, address system with fire protection in important building, auditorium, exhibition, public meeting/rallies, mobile address etc.

In most of the building the sound system is used for address by a speaker at some centralized placed or in either open area or indoor e.g. Hall, verandah or platform. In some other building like places of worship this system is also used for a program by choir and play back of some recorded sound.

So an amplifier with some input channel for one or two or more necessary microphone and for C.D. player is used. Horn type speaker are used for open area (outdoor) and column speaker for indoor hall or bigger room, while in small building or places of small height the ceiling mount speaker are used e.g. gallery verandah and platform. Box type speaker can also be used for outdoor.

If number of speaker are large their power amplifier with more zone output are used. If at some places, when some address system is needed only for outdoor persons and address for hall, for different offices, workshop, classes is independent of outside program, two separate amplifier can be used. Some good example with schematic diagram is given below.

4.14 Public Address System for a School:

The basic objectives of this sound system are

1. To conduct morning assemble or prayer.
2. Paging or announcement from principal office to the class room.
3. Reinforcement of sound in auditorium/hall.

We shall discuss the sound reinforcement in auditorium later on. Morning assembly or prayer is performed outdoor courtyard. For this single zone amplifier with some (2 or 3) microphones are used for speech and prayer. Unidirectional dynamic microphones on stand or podium are used for speech/prayer of head or group of head, while cordless microphone with there rover/receiver are used to cover prayer or speech where provision of wired microphone is not done or not possible , Four PA horn loud speakers are used for 150 ft x 120 ft courtyard for bigger field more numbers horn L.S. can be used.

The purpose of the system is to broadcast massage, announcement from the principal office to all class room. A paging microphone is used in principal office for control call attention, announcement. Wall speakers are used in each class room, staff room, offices etc. As different level of sound is recommended for outdoor and indoor rooms multi zonal amplifier is used in addition to the amplifier for outdoor assembly etc. A schematic diagram is shown below.

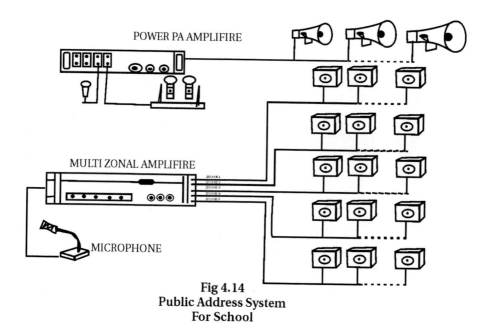

Fig 4.14
Public Address System
For School

4.15 Places of Worship Reinforcment System:

The purpose is to address by Priest/Imam/Granthi both indoor and outdoor or choir for chorus program (Sabad Kirtan, Bhajan, Kirtan, Prayer, Arti etc.) The schematic diagram is shown below.

The priest must use gooseneck microphone for preaching, the clip wired or wireless microphone for conducting prayer or program, while a dynamic unidirectional microphone for choir program or Bhajan, Kirtan etc. Mixer & Splitter can be used if more microphone are being used, as incase of a Church. As different level of sound is required in side or outside the places of worship so two zone amplifiers are used with horn speaker for outside coverage & column speakers for indoor via these.

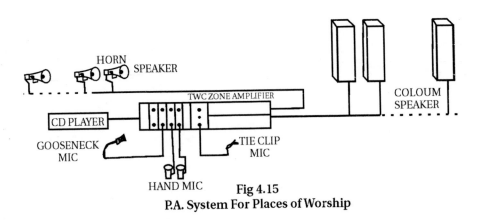

Fig 4.15
P.A. System For Places of Worship

4.16 Conference System:

The objective of this system are :

1. To make conference sound audible to all delegates.
2. To minute or record the proceeding.
3. For coverage of sound in the conference room away from table and even outside the room if required.
4. To facilitate observer not on conference table to participate in conference by using roving cordless microphone.
5. To make provision for a presenter podium microphone to be part of conference system.
6. To provide the chairman priority control over the proceeding.
7. Head phone listening for the chairman and delegates.
8. To play back the recorded sound through conference system.

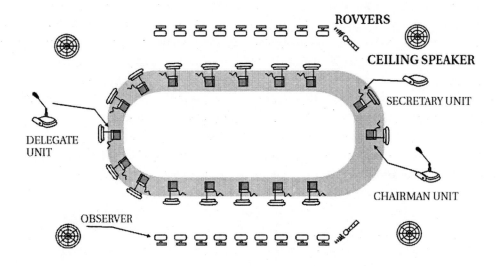

FIG 4.16A

To achieve the above objective conference sound system consist of following various unit.

4.16.1 Chairman Unit: This is a compact die cast aluminum body housing consisting of sleek gooseneck microphone with indicator for ON position with amplifier and a speaker. This unit has provision to provide head phone for more attentive listening. The chairman unit also has a priority switch beside a talk switch, which is used to interrupt a delegate at any time by muting his microphone to control the proceeding of the meeting.

4.16.2 Delegate Unit: This is similar to chairman unit except that it has no priority function.

4.16.3 Secretary Unit: This unit has speaker and amplifier with head phone arrangement only. This unit can be kept in conference room or in a separate room to take down the minutes of the meeting. This unit has additional facility for connecting cassette for playback or recording of massages and proceeding.

4.16.4 Central Amplifier:- his unit provides supply voltage to all units in conference system and control overall sound level through all units. This amplifier has provision of connecting roving and podium/presenter microphone. This also provides output for driving additional reinforcement speakers. These speakers may be ceiling mounted in conference room and wall mount or box type speaker to be placed outside in gallery or other rooms as required.

The amplifier may has provision to connect processor like graphic equalizer or feedback destroyer. The amplifier has provision for the automatic switching OFF microphone other than of delegates and chairman. The central amplifier has also the facility to interface another central amplifier for expansion of conference system.

The schematic diagram is shown in figure 4.16. The conference often has a oval shaped table to accommodate delegates with additional chairs near the wall all around the table, for observer or staff. The secretary unit may be on the table near the chairman unit or in a separate room. A normal conference system consist of 17 delegates including chairman & secretary. All the system units are placed on conference table and are inter connected in daisy chain and connected in closed loop to central amplifier. The passive participants or observer may participate in proceedings through roving microphone sitting on wall side chair.

The conference rooms prone to feedback and howling are provided with feedback destroyer connected to central amplifier with jack (Send and Return) and socket. For larger conference rooms larger delegates units are used with D.C. injection boxes/ cable to give additional supply voltage in the middle of daisy chain to compensate supply voltage attenuation.

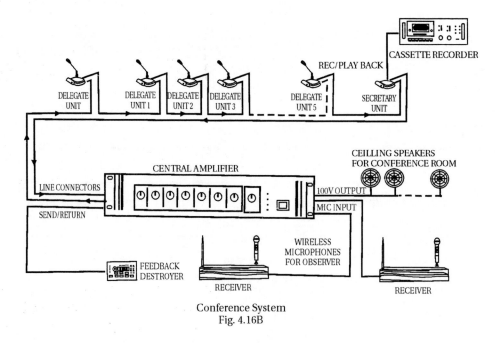

Conference System
Fig. 4.16B

4.17 Musical System:

There are mainly three type of musical sound system required for various buildings.

1. Fore ground music (Recorded Music) system in the building like coffee house, bar, lounge, fact food center or restaurant.
2. Background music (Recorded music) with paging in department store or mail or various factory or BPO center, railway station, air port etc.
3. Live musical concert or performance with paging in auditorium or studies (Film, television or radios transmission station) or recording center.

4.18 Fore Ground Musical Sound System:

Its main objective is to play back of recorded music at loud level with hi-fidelity. The system may have facility of announcement. The following equipments are used.

A stereo cassettes recorder/C.D. or DVD player suitable for commercial and home (or car) application and have built-in amplifier. These recorder or player may have auto reverse and AM/FM tuner facility. The selection of amplifier depends on the total number and total power required of loud speaker installed. A two or more zone amplifier is also used in conjunction with C.D. player.

The two way compact wall mount speakers are used with easy impendence matching. The powered subwoofers are also used for bass response. The schematic diagram is shown is figure 4.18. To provide facility for announcement the amplifier must have one or two microphone channel to provide connection for microphone.

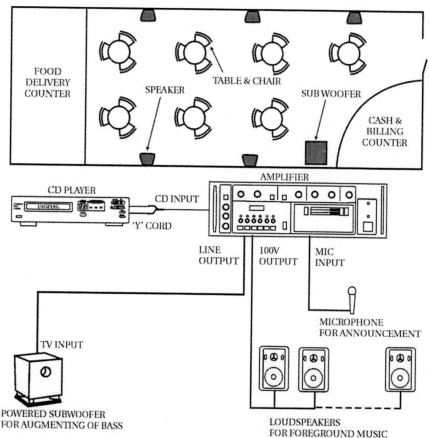

Fig. 4.18A Fast Food Centre
Fore Ground Musical System

4.19 Paging Background Music System:

Its main objective is to play back of soft music at desired level at all location and paging from main counter/Reception or central office. It may also provide paging through EPABX. System must have voice priority over riding music. To achieve these objectives the following equipment are used.

(i)An auto reverse CD or Cassette player (ii)AM/FM tuner (iii)an amplifier. System may use two channel power amplifiers to match sound level or gain of distant loudspeakers suitable for CD player and tuner input and with one or two microphone input channel. A good paging microphone is used for general announcement and emergency massage. The amplifier must also have arrangement of emergency signal or central call with voice over riding music facilities i.e. auto switching off music if voice/call/emergency is played. The amplifier may have two or more zone output to feed number of loudspeakers in different calling needing different sound level. The speakers may be ceiling mounted in small room or cabinet or wall mount in lounge, big room or passage depending upon the height of ceiling. The schematic diagram is shown below for this type of sound system Fig 4.19.

4.20 Auditortum Sound Reinforement System:

The basic objective of this system is to reinforce live program e.g. music concerts, dramas, plays or speeches etc. it also requires to maintain precise sound level for different program e.g. conference, classical or Gazal concerts or Pop or Rock music concerts. In auditorium (The word is derived form audible) must have sound effect system and to monitor system on stage and green room, i.e. paging system. Some times simultaneous translation system (S.T.) is required as in case of international conferences. Auditorium also has intercom production system and audience recall or lobby latecomer etc.

A large number and variety of microphones are needed for a music concert and vocal music. For drama or plays distance milking is provided. These all wired microphone are provided with provision of receptacles or plugs (male or female) on the floor of stage. The microphone line run from these receptacles to sound control room. The distance microphone for play may be placed near to floor to avoid peek response due to phase interference from door and reflection.

For performing vocalist a hand dynamic wired microphone is used, but for stage movement wire less microphone can be used. similarly for hand free application collar or tie clip microphone are used for actor. For lecture or speech podium gooseneck microphone are used. The hand microphone can be used on stand at stage for speech or mono act play. The provision of wire and wireless microphone must also be provided on stage for guitar or electric piano performer.

In a good sound system the selection of right type of loudspeaker at right location is of much importance. The main house speakers may be either on sides of proscenium (Stage opening)

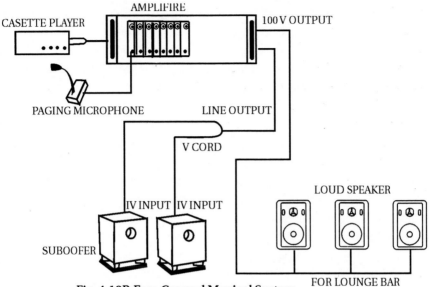

Fig. 4.18B Fore Ground Musical System

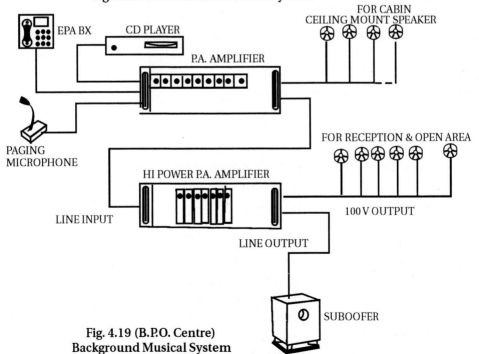

Fig. 4.19 (B.P.O. Centre)
Background Musical System

or in center in cluster. Center loud speaker in cluster is rendered invisible by sound transparent screen. There may be more than one center cluster in a row. For a distributed loudspeaking ceiling mount loudspeaker with time delay system is placed under balcony or wall mount type speakers are provided on the back of seat or rear of the hall. The horn type speakers for open area and column type speaker in lobby or lounge are used. Subwoofers are also used for bass function. For stage monitoring i.e. to know the performer on stage, to hear themself, to make suitable change in there performance and allowing stage coverage for lectures/speeches in case of large seating on stage, the wedge shaped floor monitoring speakers are used on stage to avoid feedback from adjacent microphone to avoid howling.

A missing desk with multichannel (Atleast 16 channel) is required for mixing sound from different sources i.e. from different microphone, desk, C.D. player or electronic instrument. A digital effect unit is also used with mixer for external sound effect required by performer. An automatic feed back destroyer, compressor/limiter are also used to filter frequencies to avoid howling and to control dynamic range of programs. The sound control room must be such that the operator must have unhindered view of stage.

Suitable number of amplifiers with stand by units are housed in a rack wired for monitoring and test of installed equipments. Change over to stand by equipments should be quick and easy.

Sound effect system are used to heighten illusion of realism by providing back ground music related to the scene on stage by providing sound effect speaker on stage or anywhere in auditorium.

For big convention or congress consisting of number of speakers either on stage or in hall having microphones or delegate units and speakers are in the ratio of 1:1 or 1:2. Such system must be operator/anchor controlled to ON or OFF the speakers or delegate units or with call request switch on a panel. Operator may be computer controlled for switching microphones. In an international convention or having multi linguistic delegates, the simultaneous translation or interpretation system is installed. In wireless or wired system a signal taken from main house reinforcement system to several translator booth via head phone. The translator speaks into a microphone placed before him and each of which feed one wired translation channel. Each seat has a jack for head set with a selector switch to select particular translation channel and volume control.

An intercom system is provided for communication between stage manager and sound reinforcement operator, sound effect system controller, stage light operator or lighting control booth. All-Talk and All-Listen type single loop telephone communication system is used. Wireless communication system consisting of a hand gear (single head phone and a microphone) transmitting on a same frequency can be used.

For the auditorium staff at their different working position to hear a production or rehearsal, artists in green room and paging message to them, a paging close speaking microphone with

REINFORCEMENT
AUDITORIUM SOUND SYSTEM

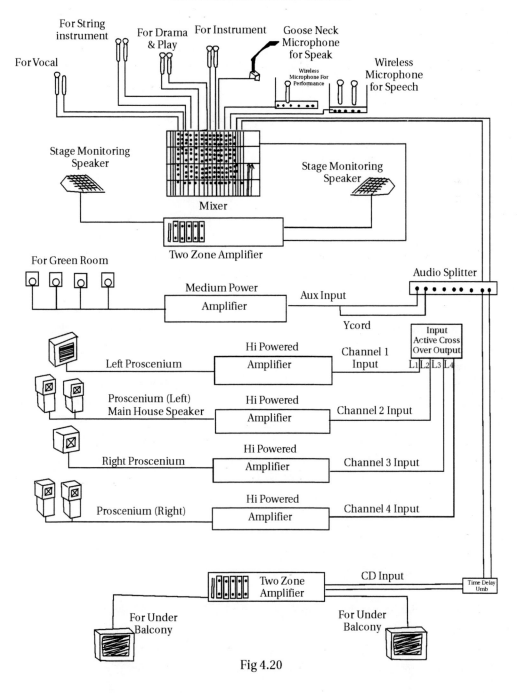

Fig 4.20

press to talk type operation are provided in wings for stage, green room, house manager room, operator room, light and sound control room or wherever required. A monitor speaker is also provided everywhere.

Provision is also made in auditorium to stop and holding attention of latecomer interrupting performance to ensure their arrival in time. The system is also provided background music during intermission beside also recalling audience. A chime or electronic bell can be used to sound audience.

A block and schematic diagram is shown in figure 4.20

4.21 Estimating & Costing:

After selection and decision of sound system to be installed in building as per requirement, we design for microphones, amplifier and speakers i.e. we decide the numbers of various type of microphone, omni directional or cardiod or super cardiod or dynamic or condenser type considering their application and need of power or operational characteristic, impedance sensibility, Similarly we design for various loud speakers i.e. horn, ceiling/wall mount or column/box loud speakers as per requirement. Amplifiers are selected as per number of input channel and out put channel required and required power. Prior to design be select the type of sound system required in the building.

4.21.1 Selection of Sound System: The purpose and use of building will decide the type of sound system to be selected for installation in buildings among various systems discussed in 4.5 of this chapter. Portion of the building or premises may have one or more type of sound system.

4.21.2 Design of Microphone: The microphone is selected as per kind of application (vocal or instrumental) and place of application, size, cordless or with cord, hand held or stand mount or desk mounted etc.

(i) Application i.e. vocal, instrument, public address, lecture, conference.
(ii) Requirement i.e. Omni or unidirectional, cardiod or super cardiod.
(iii)Place of application hand held table/floor stand mount desk or podium mounted.
(iv)Size i.e. dynamic or condenser.
(v) Cord i.e. cordless or with cord.

After consulting with user of buildings the number of microphone of different type as per above selection basis are worked out.

4.21.3 Design of Loud Speaker: The loud speakers selected out of various type of loud speakers as per their application, place of application and type of mounting.

(i) Various application i.e. public addresses, musical performance (back and fore ground music) paging conference live speech and prayer at place of worship.

(ii) Place of working, open area or ground hall or auditorium, stage program, stage monitoring etc.

(iii) Mount i.e. wall or ceiling mount

For public address in open area horn type loud speakers are used while for live speech or musical programs in a hall or auditorium following type of loud speakers are used as per location.

(a) In gallery or under balcony or in false ceiling, ceiling mounted loud speakers are used.

(b) In hall of height more than 25 ft. wall mounted loud speakers are used.

(c) In auditorium or hall with stage.

      On   (i) Front wall column type loud speaker are used.

                       (ii) Side wall box wall mounted loud speaker are used.

      and for   (i) Sound reinforcement column type loud speakers are used.

                       (ii) Musical concert column or wall mount loud speakers are used.

(d) In conference room wall mount loud speakers are used as per height of ceiling.

After consulting with user of buildings the number of different type of speakers as per above design and selection basis are worked out.

4.21.4 Design of Amplifier: After selection the number of microphones, loud speakers and required accessories like DVD player, Mixer etc. we decide number of input channel and out zone are decided in following manner.

(i) Each microphone, accessories need one input channel except in conference system in which all microphone units are connected in series.

(ii) Each accessory like DVD player, Mixer etc. need one separate input channel.

(iii) One zone output is required for a cluster of loud speaker and the number of loud speakers in cluster is decided as per output voltage of amplifier and the distances of the loud speakers from the amplifier to withstand voltage drop.

(iv) One zone output is required for accessories like subwoofer, booster amplifier, equalizer amplifier etc.

Thus designing the number of input channel and zone output, one or more suitable amplifier of proper power is selected. A mixer is selected if numbers of microphones are used to get undistorted sound. A Booster amplifier is also selected if length of cord to loud speaker is very long A Equalizer is selected for musical live program and subwoofer is selected for fore and back ground music. DVD player or Recorder or other accessories if needed are also selected. After consulting with user of buildings the number of amplifier of different type as per above selection basis are worked out.

4.21.5 Design of Wire or Cable: Suitable size and length of the wire/cable are selected to connect various equipment and accessories with each other with the help of jack is selected and there length is worked out as per requirement at site.

4.21.6 Detail of Measurement: Once the type of microphone amplifiers and loud speakers and other accessories are selected, the quantities of these various basic components are numbered and summarized in a detail of a measurement with their location. The sum of each item in detail of measurement will give total number of these items.

4.21.7 Analysis of Rates : On the basic of manufacturer prices or quotation of authorized dealer for rates of each item proper analysis of rates of each item is prepared as per standard norms to know the rates of items as per detail of measurement.

4.21.8 Bill of Quantity: Knowing the quantities and rates of various items required to be executed, a bill of quantity on standard format is prepared having items with brief specification, their quantities, rates and amount. The total of amount column will give total cost of sound system work required for execution.

4.21.9 Abstract of Cost: Abstract of cost is also prepared as per standard norms including the expenditure of miscellaneous items, called contingencies together with the cost of sound system work and centage charges, miscellaneous item (LS), if required. The total of the abstract of the cost will give the cost of the estimate or DPR etc.

5 | Stage Lighting

5.0 Introduction:

It is a special branch of illumination, which deals with providing huge light in the auditorium for general purpose, videography etc. Apart from the special flood light, beam light are required to profile, focus or highlight any person, object or program being seated, situated or performed on stage or to give special effect or contrast effect to give good impression on viewer about the program. These light are of special type i.e. flood/spot/focus beam type light with lens fitted with lamp. Normally halogen or metal halide or special lamps are used.

In ancient time cluster of focus light fitted on an board or frame etc. were used on stage and were kept in front of stage. Focusing the person or program on stage, some focus light were used to keep on side of stage to put color paper on light manually to give color/contrast or special effect.

Now-a-days auto dimmer/control panel operated lights are used on stage, which can be controlled by an operator sitting in projector room, where projector can also be kept to show video film etc in the auditorium. Auto dimmer or control panel operates all spot light, multipurpose light, effect light, scanner light etc for stage program its videography or photography, must be clearly visible by any distant situated viewer.

5.1 Various Kind of Light:

Now we shall discus the various light used, their purpose, their internal construction or composition and nature. The various kind of light used in side of stage of auditorium are as below.

5.1.1 Profile Light: This light is used for profile or high light the particular object or person. This is spot light in which light comes out in the form of beam because of condenser lens used. PT1011 profile light uses 115mm × 145mm while PT2011A or B uses 145mm × 200mm condenser lamp. It has four edge with iris shutter. T-11, 1KW or CP 43, 2KW lamps are used with GX 9.5 or GY16 lamp holder. The beam formed by this light may be round, square or rectangular, 1KW light in the called soft light. This fitting is available in black finish only.

5.1.2 Plano Convex (PC) Light: This light also forms spot of light with the help of a Plano-convex lens. Size of spot can be changed as per need and used. This light can be used for multipurpose. This light has single condenser PC lens of 115mm, 145mm or 200mm in PT -

1011M, PT-2011 and PT2001 with 1000 watt T-11 lamp and GX9.5 holder and 2000W CP43 lamp & GY16 holder. This light is supplied in black finish, black mat finish or blue finish. This is hard light.

5.1.3 Fresnel Light: This light with fresnel lens gives uniform light. It again gives scattered spot light and used for film or TV studio and stage show. These fitting PT1001, PT2001 A or B has 150mm, 200mm or 250mm fresnel lens GY9.5, CY16, G18 lamp holder and T-11, CP43 2KW and CP41 2KW lamp barn door. This light is supplied, black finish, mat finish or blue finish.

5.1.4 Par Light: This fitting has shield beam lamp. To make the colorful light gelatin paper of different color paper is fitted on cover of fitting after lens. These are available with 1KW, 300W, 70/120W or 12V/30W lamp as PAR 64, PAR 56, PAR38, PAR 36, 1000 watt lamp is used to give wide, medium and narrow beam. These PAR light are placed on side ladder made of 55mm by 25mm MS pipe. Side ladder are placed between side wing (curtain). So there may be three ladders on both left & right side of stage. Minimum four light are accommodated on each side ladder. These light are also called cross light. To give colorful effect front side PC and fresnel and par light on side ladder fitted with colored gelatin paper on their cover.

5.1.5 Halogen Quartz: These light are used as flood light for cyclorama or general purpose, accordingly they are named as,.

(A) Halogen Flood :- These are hard & soft flood light and used for film, TV studio or video outdoor shooting. They are fitted with R-7 holder and these are barn door and scrim black finish to hold. 1000watt or 2000watt halogen tube lamp (189mm) in BQ 100 or BQ 200 holder accordingly is called soft & hard light.

(B) Cyclorama Light :- These are used to light up back ground or cyclorama screen if used in last stage. When all lights are off, these light are thrown on screen and soft light reflected from screen are obtained on stage. They are useful to highlight banner fixed on screen. These fitting have curved aluminum finish dots reflector and R7 holder/socket to hold 1000 W/500W halogen (189mm) tube lamp in CFH100 or CFH50. These are available in black finish with color frame.

C) Cerneral Purpose Light :- They are used for general purpose flood lighting on stage. Theses light have curved aluminum finish dot reflector and 2×R7 holder/socket to hold1000W, 189mm halogen tube lamp, in 100H or 50H. These fitting are supplied in black finish Mash and Barn door.

5.1.6 Effect Light: These are used to give particular effect on stage or screen, on dance floor with D.J light. There are various type of effect light.

(i) Trafic Light: These light are in the form of traffic light. It has three or four colored light fitted in colored frame. It is operated with chaser or it has built in chaser unit to make them blinking one after other (sequential effect). It is either 4×100W (R,Y,B,G) or

3×100W(R,Y,B). It is useful with D.J. light or with dance floor for dancing or similar program.

(ii) ***Strobes:*** It gives white light. It is either fitted with frame and are electronic dimmer operated. The fitting have one or two tubes giving blinking effects. These are available with 30W & 1500W. The light is useful with DJ set so it is DJ light but it is often used on stage to give white light.

(iii) ***Scanner:*** It is a fitting which come after scanning through colored plate or pattern. These light have two steel rotating disc, having circular cut. One disc have colored glass in to circular cut, other disc has different pattern to give different color beam with different pattern on stage floor or screen. Some of light are sound active like SAKURA. Carousel. These are fully automatic, multicolored beam with sufficient gobos e.g. AUTOSCAN has 12 gobos combination with multicolored with 24V, 250W lamp and CARON has 4 gobos only. Some other common scanner light are PRISCAN, MINISCAN, MONOFOLOWER etc. These are useful light for dance & similar stage program.

(iv) ***Galaxy 12"150W :*** These are supported with 15V 150W lamp and gives multi colored beam. This effect light is useful for Disco and Disco parties.

(v) ***CRAZY 500W:*** These effect light have RJ 7s holders to hold 500W halogen tube to give muticolord beam. This light is also used in Disco or Dance parties.

5.1.7 Sound Active Light: These light operate with particular frequency of sound. So these light are fitted with sound active plate to switch on the light. These are used with D.J. and known as D.J. light. They are some time mounted on dancing floor. These light have 12V 250W one or two lamp. Very common sound active light are MAGIC I or MAGIC III, MARK 250. Some of scanner light are also sound active.

5.1.8 Cool Light: These light uses 50W lamp, which are daylight or tungsten. They gives soothing effect on eyes and known as cool light. These light have 2,4 or 5, 55W CFL lamps with mirror finish reflector to give light in the form of beam. These fitting are available with simple. ON/OFF switch or operated with analog dimmer or DMX panel. These are supplied in black finish commonly.

5.1.9 H.M.I. Light: These light uses metal halide lamp or luminair which gives day light illumination with high light output (lumen) at low amperage and consistent color balance. The HMI light is a spot light with a versatile daylight source, which is extensively used is motion picture or film studio, video production & out door location e.g. HMI4000W or HMI2500W, HMI1200W, HMI750W are also excellent daylight spot light with less out put and used in film/television studio outdoor shooting and news coverage. They work as fill light and daylight booster HMI gives full range spot to flood.

5.1.10 Solar Light: These are versatile fresnel spot light with 50mm/250mm convex lenc and spherical, aluminum anodized reflector and may use 1KW or 2KW lamp with GX 9.5 or CP22 or CP938 socket. These steel housed fitting are used as key back or fill light.

5.1.11 F.O.H. Light: These are number of lights in the form of a house or cluster fitted is front of stage to give sufficient light on stage as per size of stage and requirement on stage to cover every place on stage. These light are called front of house simply FOH light.

Some important equipment used for operating different above lights are mentioned below.

5.1.12 Chaser: This has an arrangement which connect the supply to the different circuit, which has light (stage light) to give blinking effect. These may be multi or sequential type to switch on light in sequence or multi light in blinking mode. Some chasers are musical chaser with sound active plate.

5.1.13 Smoke Machine: This gives smoke or mist effect on stage in which effect light become very impressive & effective. These electronic machine, which gives jet of thick chemical which appears as smoke from distance. This equipment is in black finish.

5.2 Fitting of Light:

We shall discuss about how these stage light are to be fixed and how we work out number of light required.

5.2.1 F.O.H. Light: As we have discussed the F.O.H. light gives light on stage from front side F.O.H. light can be fixed in two manner.

(i) *On Light Bar:* A bar is provided in front of the stage at least seven meter from the stage in a standard size auditorium made of 50mm×25mm 14/16 SWG MS pipe. The bar is suspended from false ceiling support or trust about 60cm below ceiling. False ceiling is made curved in such manner so that light must not visible from back or sitting person in auditorium. This bar should not just above the sitting person to avoid fall of broken lense on the sitting person. Lenses often crack& breaks due to expansion at 1800°C temperature and contraction when light is switched off. Broken hot glass may cause fatal damage to any human being on stage or in hall falling from top.

The bar should be of sufficient length to accommodate at least 16 lights. 8 of these light are pointed to the left from centre & while remaining 8 towards right to cover light on whole stage. We can use or lit light as much as required to lighten any part or complete stage. These light may be 1KW PC (Plano Convex) light. Out of eight lights on each side one of there may be profile light to high light or focus any person.

(ii) *On Side Rift:* F.O.H. light can also be provided on side walls instead of suspending them from ceiling. This is better arrangement to focus light from two side wall i.e. left or right accordingly they are called left or right rift. The rift is again made of 50mm × 25mm 14/16 SW MS pipe. These are made in eight section to hold eight light i.e. 16 light in all on both rift as in case of light bar. These lights on rift are 2KW PC light in stead of 1KW as in case of bar. We use one profile light on both rift out of eight to high light person as in case of bar light arrangement. These rifts have an arrangement to rotate the rift about 120° with help of bust used on two brackets. The light in this arrangement has better chances to reach every part of stage to cover full stage.

5.2.2 Other Internal Light: To fix other stage lights inside the stage a steel super structure is constructed on stage below the truss of auditorium. All stage light together with automatic stage curtain multimedia projector system etc. are fixed on the super structure. This is also very useful for provision of sound system on stage. For both construction & maintenance a cat walk is made of steel and wooden board just above the steel super structure. for the movement of staff

To construct the steel super structure it is essential to study the length width and depth of stage. The stage is of sufficient depth so the stage is divided into three section or part. Where three stage are made, namely front, middle and rear. They have separate automatic curtain, frill & side wings. We use last or rear part for effect light on effect bar & a cyclorama screen while first & second stage are main stage which has first & second light bar on which profile, fresnel, PC, halogen are fitted. The first bar normally have 21 light namely, 3 profile, 8 fresnel, 8PC and 2 halogen, while 2nd bar on middle stage has eighteen light i.e. 2 profile 7 fresnel, 7 PC & 2 Halogen, we have already discussed that first & second bar may have more than 21 or 18 light as per requirement and size of stage. Third effect light bar in rear stage have minimum eleven lights for cyclorama. These may be more as per size of screen (cyclorama). The effect bar has minimum 2 scanner, 2 strobe, 2 cyclorama and color lights are fixed as per requirement. Cool light, HMI light & DMX light are used as per requirement on stage.

Par light as discussed earlier are fixed on side ladder fixed in between wings on each stage. These bars (first, second and effect) and side ladder are suspended or fixed with steel super structure. Steel structure is made of 50mm MS round pipe and fixed on three wall of stage. i.e. two side & one rear wall, so that load of super steel structure is distributed uniformly on three wall. This structure is tied with truss of auditorium at different places to avoid vibration due to walking on cat walk and for more stability. The cat walk is made of suitable size of angle iron and wooden board.

Total load on steel structure is first calculated before designing it i.e. the weight of all lights &three curtain and frill. The curtain has 165Kg weight, but during shifting it has 180 kg because of tension. So in all each curtain has 180+165=365kg and three curtain has 3×365=1095kg. Total load, including weight of light, curtain & cat walk usually comes out around 2 ton on super structure. So superstructure is constructed at sufficient height, approximately 7 meter from stage level.

Normally no false ceiling is made on stage. Although auditorium seating hall, usually have false ceiling to give aesthetic look.

If false ceiling is used on stage it is made about 1.8 meter above catwalk to facilitate movement of maintenance person. Two exhaust fans 24" or 18" are used to exhaust the heat due to light on stage in side false ceiling.

5.3 Wiring and Operation:

After design & selection of different light on 1st, 2nd and effect bar in stage, FOH rift we energize these light and operate them as per our use. Most of light are of 1KW or 2KW and few are of 500W & 250W so separate circuit is used for individual light with 2 core cable or 2 wire single core to energize in parallel with 3 phase distribution at patch panel. Common earth wire can be used for each light circuit with patch panel by loop method. As total load of all lights are above 5KW, so three phase balance distribution is used. Every light is connected to patch panel, which may have 96 or 124 point to connect equal number of light. So a patch panel works as a 3 phase distribution box connected to main control TPN as discussed in chapter of writing & distribution box.. Each light is controlled individually with SPNMCB. Wiring is done in PVC pipe or cable tray or channel for easy maintenance. MCB normally kept ON while operating stage light through dimmer switch on dimmer panel but MCB is useful to work for maintenance to put circuit off to cut the power or not to operate specific light, as they are connected to the panel of dimmer control and is grouped in MBCB box. A channel is usually made of 4KW but it is advisable to use it on under load say maximum 3KW. So depending on the load of different light, number of light circuit is connected to one channel through dimmer switch and indicating bulb. The dimmer may be logarithmic DMX or analog signal electronically controlled. In analog dimmer switch is operated manually, while in digital DMX is controlled by computer compatible signal. Normally we do not use digital DMX dimmer as it has fixed time bound program, which many not match with running stage program & may operate stage light prior or delayed with stage program.

To operate stage light we have a controller in addition to dimmer control panel, which can be placed and operated from any place, while dimmer panel is placed at fixed place. Controller is a sort of mixer, which give common/control signal and have both digital or analog signal. Both Dimmer & controller may have 6 channel, 12 channel or 18 channel or 24 channel according to the number of stage light to be controlled. We can put some light off by patch panel by removing lead of light from patch panel.

Dimmer may be mechanical with auto-transformer or electronic with log type transformer. Much more technical advancement in sound & light industry has resulted increased application of electronics and computer in today's entertainment arena and resulted in perfection in entertainment technology. Many firm has come up with various world class range of product. DMX 512 dimmer, named Power Station and DMX to digital analog converter. Output analog voltage in converter can be given to 24 channels. D-25 type output connector is used to give output to analog dimmer, while XLR 3 pin connecter gives DMX output. Each master fader, output has LED indication of various color.

5.4 Precaution:

Normally we use less then 10 light or light point or maximum 10 point with electronic dimmer. There is a fluctuation of voltage and excessive heat results in fuse of lamps. So we use digital volt meter/multi meter to watch voltage fluctuation. Imported MH or halogen or UVI bulb are used as stage light bulbs are not made in India as it require 32°C quartz. These bulbs are manufacturing in cold countries. These lamps can not with stand voltage fluctuation. The voltage must be between 210 volt - 220 volt and not above 240 V. High voltage or low voltage results in excessive current and may fuse these lamps. Japan has established a lamp factory below ground under sea. An imported bulb can give radiation at 3200°K in color spectrum. While in India we manufacture halogen bulb which give light rays at 1800°K So imported halogen give very bright light and is very useful for photography.

Dust must be removed form each light fitting before use by air blower, as it is deposited on fitting. Due to continuous sufficient use, the light fitting gets heated resulting in burning of dust creating smoke which is sensed by smoke detector to give false signal. Normally light fitting are dust vermin proof so the dust is not allowed to deposit on bulb especially to result excessive heating or breaking of bulb. Similarly dust is also not allowed to deposit on electrical dimmer or controller. So dust do not penetrate inside and deposit on carbon, which may result in sparking causing damage to panel or disturbance in sound system. So sound & electric cabin/rooms are kept separately, if possible.

Light are used as per programmer. All lights are not used unnecessary F.O.H. lights are used in every program. Number of F.O.H. light used only as per requirement. Similarly when effect light are used, all other than color light are put off. However selected effect lights are used is rotation changing one by other to give better effect and good program & avoiding unnecessary use of more light at a time.

5.5 Estimating & Costing:

To prepare estimate & to evaluate the required cost, we divide the stage into number of sub-stages, 2 or 3 as per of size for making preparation of stage program.

5.5.1 Design of GI Pipe: As 50mm B class GI pipe to construct first, second or effect stage bars, super structure etc. we work out the length of pipe as per size of the stage. F.O.H. bar or rift (left & right(are also need for general purpose light or spot light to high light object.

5.5.2 Design of Stage Lights: As per standard norms the Profile, Plano Convex (PC), Fresnel, Par light are provided on light bar i.e. first and second stage bar etc. The cyclorama & effect lights e.g. Scanner, Strobes, Crazy galaxy, Traffic etc,. as needed, are fixed on cyclorama or effect light bar. Some of effect light may also be fitted on light bars. The Spot & General purpose Halogen lights are provided on F.O.H. or rift while some of them are also provided light bars as discussed earlier generally 21 & 18 lights are provided on first and second light bar, 11 lights on cyclorama or effect light bar. The light may be more on these stage and effect light bars as per size of the stage as discussed in 5.2.2 of this chapter. Accordingly the numbers of various lights are counted.

5.5.3 Design of Dimmer and Control Panel: Knowing the number of different lights, Dimmer Control Panel, Patch or Cross connecting panel, Digital or Analog DMX dimmer are selected. A controller or mixer is also selected if required as per 5.3 of this chapter.

5.5.4 Design of Wiring: Wiring for connecting each lights in parallel as required, is done either two wire submain or two core unarmored cable in cable tray or channel with common earth wire as discussed in 5.3 of this chapter. Accordingly the length of submain or cable calculated.

5.5.5 Detail of Measurements: The number of each light par, PC, fresnel, Profile, Spotlight, General Purpose light, Effect Lights, Smoke Machine, Cyclorama, Screen etc., length of 50mm GI Pipe, rift of required section, control and patch panel, dimmers/controllers, length of cable are summarized in a in a table named detail of measurements.

5.5.6 Analvsis of Rates: Many manufacturer company e.g. Modern Stage Light, Sai Stage etc. are manufacturing these stage light apart from the other foreign companies exporting these material. Knowing the price list of these companies, proper analysis of rates are prepared, if there exist no schedule of rates.

5.5.7 Bill of Quantity: Knowing the quantities and rates of each items the bill of quantity is prepared on standard format to know the exact cost of installation of stage lighting.

5.5.8 Abstract of Cost: Apart from the cost of lights as per bill of quantity miscellaneous expenditure are added in shape of contingencies or any miscellaneous items as lump sum in abstract cost. The total of the abstract of cost give total size of estimate or D.P.R.

PROFILE   FOLLOW   SPOT

FRESNEL

PC

PAR

STROBES

HALOGEN

HALOGEN

STROBES

SMOKE
MACHINE

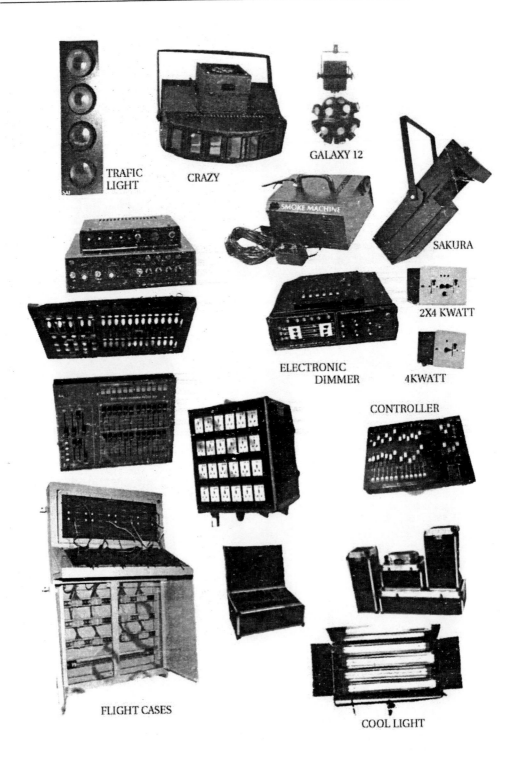

TRAFIC LIGHT

CRAZY

GALAXY 12

SAKURA

SMOKE MACHINE

2X4 KWATT

ELECTRONIC DIMMER

4KWATT

CONTROLLER

FLIGHT CASES

COOL LIGHT

AK MSR 1200W
FOLLOW SPOT

AK HMI 4000W
HMI FITTING

SOLAR 2 KW

330 WATT COOL LIGHT

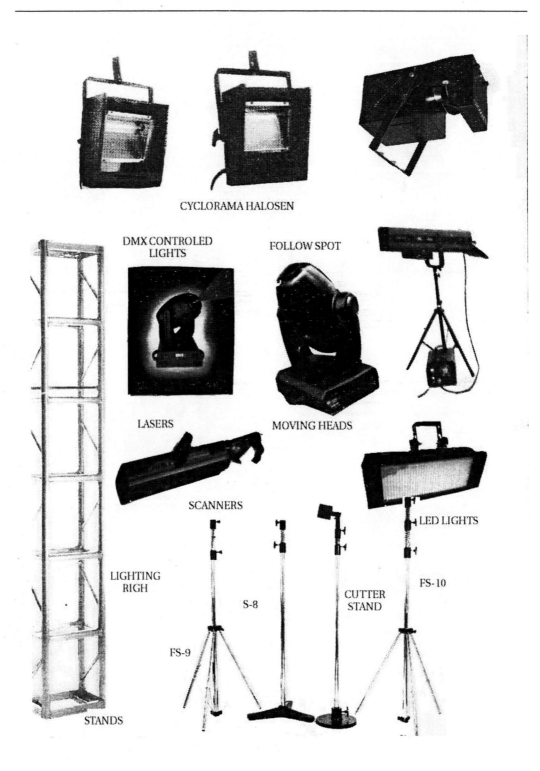

CYCLORAMA HALOSEN

DMX CONTROLED LIGHTS

FOLLOW SPOT

LASERS

MOVING HEADS

SCANNERS

LED LIGHTS

LIGHTING RIGH

S-8

CUTTER STAND

FS-10

FS-9

STANDS

6 | Diesel Generating Set

6.0 Introduction:

In early days of beginning the universe and life on earth, man was totally dependent on natural things, natural eatable to eat, water, to drink, natural light(Sun) to see each other. In the darkness the flame of burning wood was used. With growing need & invention of cotton, cotton dipped in kerosene (Kerosene lamp) were used for light and the man were source of energy to carry out work. i.e. manually or using leaver. Candle lamp, made of wax were remain common and leaver and hand machine were popular for long days. The population was confined to live & survive near natural source of water. With the further growth of population certain machine were developed to do specific work driven by fall of water on impeller or driven by round movement of bulls with wheel and shaft to get rotation. Till then above mentioned source of light were prevailed. The people invented batteries and with gang of batteries were used to glow light lamp and to do work. A scientist Johan Flaming invented generation of electricity. D.C. and A.C. generator came into existence. The people constructed large hydro-power station, thermal power station to generate electricity as artificial source of energy.

Fleming right hand rules says that by rotating a coil in magnetic filed e.m.f. or voltage is generated in rotating coil. Hydro or steam turbines were used as prime mover. Further with the invention of I.C. engine, engine driven machine were used. To cope the demand of increasing population, with use of electric motor, many machine and appliance came into existence. Power or electricity has become backbone of mankind so that it has become essential service like water. The generation of electricity has fallen short to cope with in increasing demand due to short of natural material needed for generation, like water using in hydro power house, coal utilized in thermal power station and due to various maintenance problem of generation. So it has become essential to use your own generator to meet the requirement of electricity.

In the large building, it has become essential part of electrical installation to use Diesel generating set to run various electrical installation in the absence of electricity supplied by power corporation, a supplier, as mentioned in electricity act & rules.

With increase of gen-sets, sound pollution has increased in recent past. So the government has banned gen-sets making noise and made mandatory a criminal & punishable offence to manufacture, sell & use of gen-sets with noise.

So it is essential to use a soundless gen-sets in a building. The scope of this chapter is to impart certain knowledge of D.G. set, their installation, use and maintenance.

6.1 Generation Principle:

A British scientist John Phritz Fleming invented generator & he named its principle of operation, as Flemings right hand rule.

According to this rule, if the motion is represented by thumb of a right hand and first or pointing finger represent magnetic field keeping thumb, first, second finger at right angle to each other the second or middle finger represent direction of flow of current.

So if a coil is suspended in a magnetic field acting to the perpendicular to the coil and the coil is rotated such that direction of motion is at right angle to field, then e.m.f. is generated in the coil, so the current flows in the coil if connected to load in the direction, as per Fleming right hand rule for production of high voltage, the magnetic filed is developed by electromagnet formed by exciting field coil in magnetic pole.

6.2 Classification of D.G. Set:

The gen-set may be a single phase or 3 phase D.G. set 1 phase gen-set are available from 2KVA to 30 KVA. In general there are of lower capacity. Some special single phase diesel generator of high capacity 50 KVA are also available for special requirement. A single winding of generating coil single excitation field coil are used in single phase generator. While 3 winding at 120° apart on a circular rotor and 3 load or generating winding at 120° part on stator are used in 3 phase generator such that 3 phase voltage at 120 phase difference are produced in 3 winding connected in star.

Gen set of lower capacity less than 2 KVA are petrol engine driven. Some of them are petrol run while some are petrol start kerosene run. As petrol is costly and due to invention of invertors, these are not much in use, however they are useful in remote area where there is no electricity to charge the invertors battery.

3 phase diesel generator, common in use for large capacity of load requirement are available from 5 KVA to 2000 KVA. The Gen-sets can also be classified on the basis of type of alternator used. There two type of alternator 1 slip-ring brush or revolving armature filed(2) Brush-less AVR AC generator.

Generator may also be classified on the basis of the prime mover i.e. engine used, engine may be water cooled or air cooled and self aspirated or turbo charged.

6.2.1 Revolving Armature & Stator Filed Alternator: In this D.G. set field is static and mounted on salient pole stator yoke. The stator frame is made of cast steel or fabricated with electrical steel having yoke for magnetic circuit. Its coil is fed from self exciting and self regulating exciter unit and rectifier as shown in figure i.e. exciter unit is fed from alternator output. Initially it is operated on residual magnetism voltage of 12V battery. Its output is taking from armature wound on rotor through slip-ring and brushes and terminal box. The rotor is rotated in bearings fitted in stator frame.

Prime mover is a diesel engine, water or air cooled with or with out TC, and is usually operated on 1500rpm to give rated output.

6.2.2 Brushless Alternator: In this type of D.G. set field coil is rotating and the output is obtained at stator winding where armature is wound. The stator is made of high quality high permeability low loss steel, but rigid to withstand any kind of forces. The armature coils are made of copper with H class of insulation with 2/3 pitch winding to eliminate any harmonics generated in output voltages. Windings are double dip impregnated and its overhangs are provided with epoxy gel coating to protect against dust and corrosion. Twelve winding leads itself form output terminals from armature without joints, which are sleeved and terminated with crimping socket in a terminal box.

The rotor is made of high permeability low loss steel stamping built on rotor shaft. Rotor is wound with field coil of copper with H class of insulation. The exciter is in two part (a) Exciter stator made of high quality laminated steel stamping carrying field coil mounted on non drive end shield of machine. The copper coil of stator exciter as H class of insulation, the exciter rotor built on iron insulation in the form of riveted core carries 3 phase armature winding mounted on shaft. The rotor rectifier winding is connected to rotating rectifier assembly, which is 3 phase bridge rectifier and is connected to field coil of alternator on rotor. The exciter have two exciting disc mounted on rotor. The rotating part is made dynamic insulation hub to hold diode of rectifier.

The AVR is compact unit mounted inside terminal box. The unit senses the voltage between lines and controls generated output voltage, its potential meter adjust stability, frequency roll off and quadrature drop during parallel operation. It can build up from a very low residual voltage about 2.5 V (line to neutral) and response of AVR is very quick.

A block diagram of AVR is shown in figure 6.2.2. The brush-less D.G. set is self excited, the exciter has residual voltage. In its absence at the time of starting exciter field coil is connected to 12V battery disconnecting AVR.

Two space heaters (optional) suitable for 230V/24V single phase are provided on each overhang of main stator winding of generator on request.

The other common component in both type of alternator are as below :

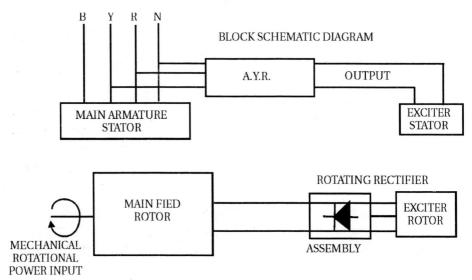

Brushless AC Generator
FIG 6.2.2

6.2.3 Bearings: D.G. sets are provided either single or double bearing. D.G. Set rotors of single bearing at non driving end (NDE) are locked at driving end (DE) to prevent damage during transit. This lock is removed before it is coupled with prime mover. Some D.G. sets are provided with two bearings, one each at NDE and DE instead of single bearing to give better control and safety and do not require any locking arrangement during transit. The bearings are lubricated with grease of proper quality and quantity as prescribed by manufacturer. Excessive grease must not be used to prevent overheating.

6.2.4 Ventilation: The AC generator are self cooled. The centrifugal aluminum fan fitted on the shaft at the driving end of rotor draws fresh air through opening provided at non driving end cover, discharge through ventilating duct in DE shield in double bearing machine and through adopter in single bearing machine.

6.2.5 Under Speed Protection: With the variation of speed of rotor load on generator or temperature, exciter field current varies through terminal voltage after AVR is constant. So under speed protection circuit is provided to save generator.

6.3 Prime Mover:

The prime mover is a diesel engine of proper HP to give required KVA of generator. The engine may be water cooled or air cooled, self aspirated or turbocharged. The engine usually operates at 1500rpm to give rated output.

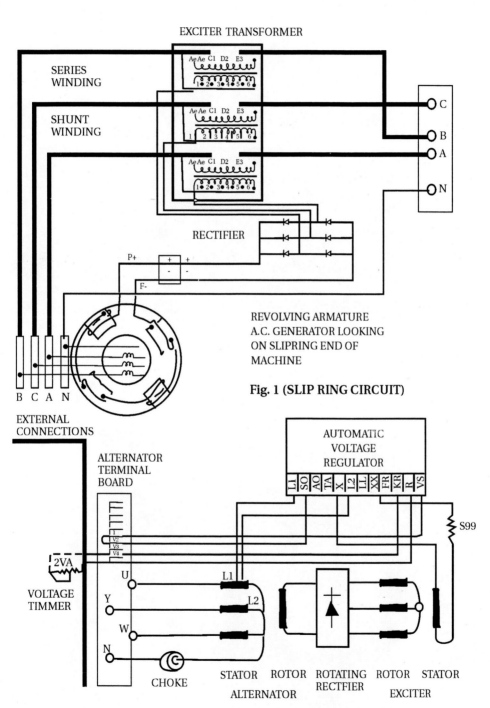

EXCITER TRANSFORMER

SERIES WINDING

SHUNT WINDING

RECTIFIER

P+

F-

REVOLVING ARMATURE
A.C. GENERATOR LOOKING
ON SLIPRING END OF
MACHINE

Fig. 1 (SLIP RING CIRCUIT)

B C A N

EXTERNAL
CONNECTIONS

ALTERNATOR
TERMINAL
BOARD

AUTOMATIC
VOLTAGE
REGULATOR

L1 SO AO TA X L2 LL XX FR KR R VS

S99

2VA

VOLTAGE
TIMMER

U

Y

W

N

CHOKE

L1

L2

STATOR ROTOR ROTATING ROTOR STATOR
RECTFIER
ALTERNATOR   EXCITER

Fig. 2 (BRUSH LESS CIRCUIT)

ENGINE

ALTERNATOR

6.4 Control Panel:

There are two type of control panel of D.G. set.

(a) Manual Panel (b) Auto manual failure AMF

6.4.1 Manual Panel: This is fabricated from sheet metal and is air cooled. Its front panel is in three section. One section has all the power output section, second has all engine control and third section has the remaining control and all indicating instrument.

safety unit (ESU) is to monitor and starts & stop the generator as necessary and is installed on panel board. This unit has engine safety & operates relay & glow related indications as per fault. There are five fault safety namely (1) Low oil pressure (2) Battery not charging (3) High water temperature (4) Auxiliary emergency stop (5) High temperature stop.

On arising fault it activate stop solenoid which stop the fuel supply vai a rod, to stop engine and glow corresponding indicating five LED lamps, out of which one LED for fuse if it blown & other four are for four faults. LED for high temperature or water temperature is same, panel has a voltmeter & and ammeter with their selector switch.

There are two more indicating lamp on panel (1) Set on lamp (Green) connected in parallel with line indicating running condition of set (2) Load on lamp (red) connected between line & neutral indicating the load i.e. flow of load current.

Oil pressure gauge (PG) and a charging ammeter are also provided on manual panel to indicate lube oil pressure and charging current of starting battery.

Apart of these all above meter & lamp ESU unit, panel has ignition key switch to start engine through starting motor, emergency stop button for emergency stop purposes and current transformer to measure line current by ammeter. An MCB is also provided in line to break supply in case of over load.

For protection of instrument meter from supply variation, fuse lines are provided in fuse case. By putting ignition key on ON /start position DS set starts & putting on Off to stop the engine in normal running condition, except in emergency then engine is stopped by emergency stop switch

6.4.2 Amf Panel: This panel is fabricated with 14/16 SWG CRCA sheet metal and is in two section. Front section consist on volt meter (V), frequency meter (HZ), Ammeter(A), DC voltmeter (VM), DC ammeter(AM) with their 3 way with off selector switch, volt mete selector switch (VSS) to measure voltage, BSS for frequency of DG/mains load & ASS for current of output from engine battery. There is one 3 pole 3 way alternator/bus/mains selector switch BSS to select terminal connected to alternator, DG or main supply or bus bar to measure frequency of DG set (alternator) main power supply from power corporation or output supply for load respectively. There is one more 2 way with off selector switch MAT to select manual/auto mode of DG set. To make battery charger ON/OFF one more selector switch BCS is also provided.

This front section of panel also consists of four push button switch namely PB₁ to start engine, PB₂ to stop engine, PB₃ ACK alarm and PB₄ is reset push button. On the back side of front section alternator (CA) and mains contactor (CM) with MCCB (4 pole control switch) is fitted. Ammeter works through a current transformer provided in panel. Apart of all these, ten indicating light LED L₁ to L₁₀ are provided on the top of panel to indicate as follows.

L1   Load on DG set

L2   Load on main

L3   Under voltage

L4   Over voltage

L5   Earth fault

L6   Over load trip

L7   Low lube oil pressure LLOP

L8   High water temperature

L9   DG set fails to start

L10 Battery charger ON

On side of panel another section consist of engines start relay ESR, alternator voltage relay AVM, under voltage monitor UVM earth fault (EF) over voltage relay (OVR) engine cooling timer T₁, cranking relay, battery charger etc. Out going cable enter from bottom of panel. AC circuit & DC circuit required to control wiring & connection to various relay, with meter contactor are shown in diagram Fig 6.4.2B. The basic function of AMF panel is to auto start or stop engine on the failure of main supply that is why it is called auto mains failure.

6.5 Installation:

A great care and precaution are taken prior to installation in alignment of engine and alternator & handling of D.G. set, so that its winding, insulation, surface of slip-ring and bearings are not damaged.

The alternator or D.G. set is kept away from moisture, acid, alkali, oil, gas, dust and harmful material and must be installed at dry and clear places. It is better to construct a generator room having good ventilation of air. D.G. set is installed on a horizontal surface at proper foundation. The alignment of engine and alternator should not be disturbed during installation, so it is rechecked after installation. Exciter unit, all connection and earthing are checked before start.

6.6 Selection of Site:

Location of gen-set or D.G. set has directly affects its performance. The generator should not be located in polluted atmosphere such as corrosive fumes, cement, dust, fibers, furnace for chemicals in radiator of the engine gets clogged in such atmosphere, and got heated in the

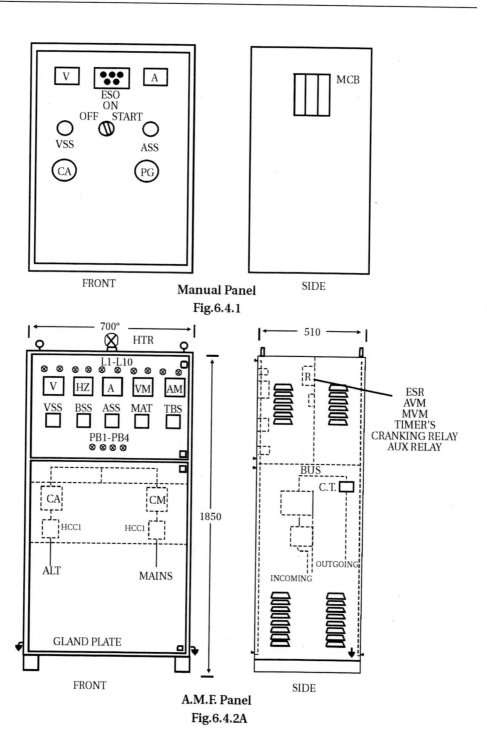

FRONT   **Manual Panel**   SIDE

Fig.6.4.1

FRONT   SIDE

A.M.F. Panel

Fig.6.4.2A

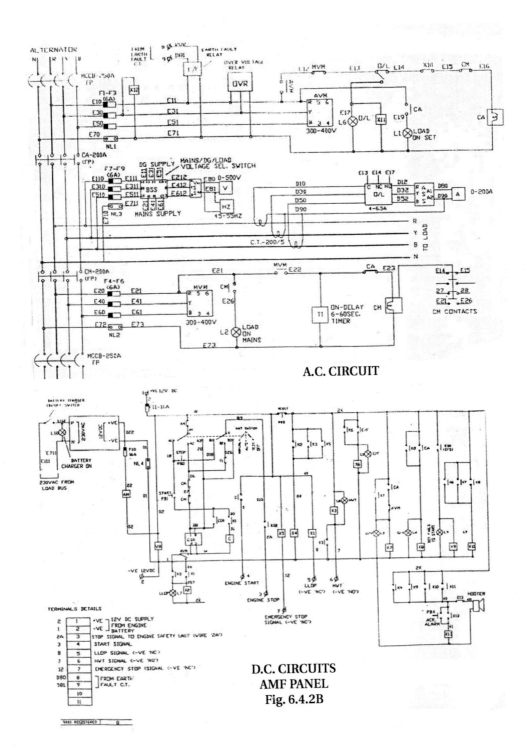

A.C. CIRCUIT

D.C. CIRCUITS
AMF PANEL
Fig. 6.4.2B

absence of its cooling. However heat exchanger cooled engine should be installed in polluted atmosphere. Frequent cleaning, once a weak, is essential to clean cooling passage by blower. Proper space should be there for placing gen-set and opening of doors. There should be 1.25 to 1.5m free space around generator for proper operating & maintenance. There must be cross ventilation of air to ensure proper natural cooling, exhaust gases & hot air discharge should not air circulated inside the acoustic enclosure (canopy) of silent generator.

D.G. set is preferably installed in open area. If AMF panel is out side acoustic panel by then a shelter with roof of sufficient height is essential for panel. If generator is to placed inside a room, there must be sufficient ventilation exhaust fan is used to discharge hot gases or air, for provision of fresh air suction. The room structure should be capable of bearing static dynamic load of gen-set and the size of room be such that their distance between wall & D.G. set is 1.25 to 1.5m & height clearance of generator from roof must be 1.8m D.G. may be placed near distribution panel. There may proper space for earthing.

The dust must not be allowed to enter the gen set room. Opening around the radiator must be sufficient say 1.5m for proper cooling.

In case of heat exchanger cooled engine is used a forced ventilation is to be provided by fresh air blower and exhaust fan to blow out hot gas or air with proper ventilation. The temperature inside room shall net be more than 5° to 10°C. For air cooled engine additional window are provided in room. Room size layout & with foundation is given in figure 6.6.

6.7 Foundation:

The foundation should be water leveled and at least 5cm above ground level. The length & width of foundation should be at least 20 to 40cm more than that of generator. RCC foundation or MS I or ISMC channel may be used to cater the load of generator. Sand filling around the rigid foundation should be done to minimize vibration. A drawing of foundation is shown in figure 6.7.

6.8 Earthing System:

Earthing system is required for the protection of equipment as per Indian electricity rule 1956, separate earthing is done for body and neutral. Double earth is needed both for body and neutral. So 4 earthing is required and distance of 2.0 meter is maintained between two earthing or earth pit. Earthings are done as per IS code. The earth resistance should not be more than $1.0\,\Omega$. The earthing electrode and conductor are selected as per rating of D.G. set. For 15 KVA to 125KVA DG set, copper strip 25×3 mm or 25×6mm GI strip is used with 600×600×3mm copper earth plate or 600x600x6mm GI plate, while 25×6mm copper or 50×6mm GI plate with same copper or GI strip for 125KVA to 1250 KVA is used.

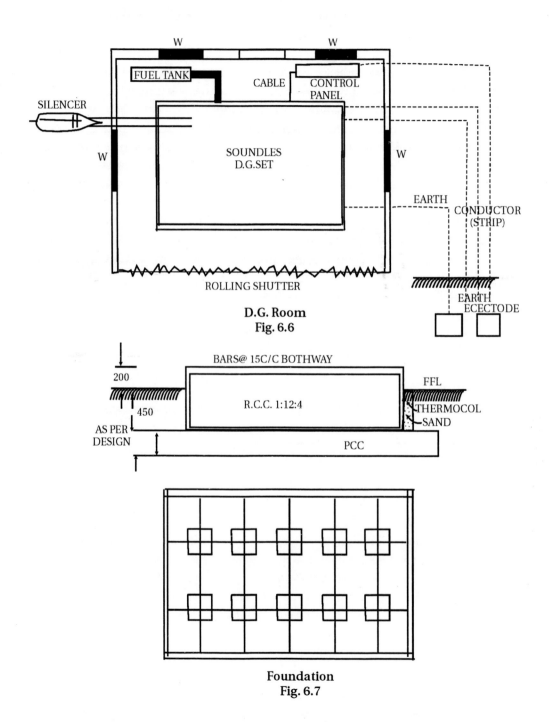

D.G. Room
Fig. 6.6

Foundation
Fig. 6.7

6.9 Exhaust Piping:

The exhaust system is so designed to keep the resistance to hot gases (back pressure) as low as possible and load of extended pipe of silencer should not be on engine manifold. M.S. exhaust pipe must be properly supported and exhaust elbow must not loaded for any stress. The exhaust pipe should be extended, if needed through proper flange and a uniform gap has to be maintained between elbow flange and exhaust pipe flange. Proper size MS pipe should only be used, while GI pipe or bends should not be used. The dia of MS pipe is selected as per capacity of DG set and type of engine used as shown below.

DESCRIPTION	AIR COOLED			WATER COOLED		R SERIES		SL SERIES	K SERIES		
RATING	15-35 KVA	40-625 KVA	75 KVA	15-35 KVA	40-625 KVA	75-825 KVA	100-125 KVA	120-200 KVA	275 KVA	320 KVA	400-1250 KVA
EXHAUST PIPE Dia mm	63	75	100	63	76	100	125	125	125	250	2x150

6.10 Fuel Piping:

A supply & return fuel pipe is required in each D.G. set. Sets upto 200 KA fuel tank is fitted inside acoustic enclosure with 12mm fuel pipe, while sets above 200 KVA, the fuel tank is to be fitted out side acoustic enclosure (canopy) with B/C MS pipe of 12 mm internal dia. and above 250KVA 19 mm-25mm ID MS pipe is recommended for DG set. MS pipe is used with proper welded joints. For out side mounted big fuel tank a stop valve and flange should be used. All joints must be seepage or leakage free. Fuel pipe should be above ground level on proper MS support. For DG set of 350 KVA or above, the height of fuel tank should be such that lowest level of fuel tank is above diesel pre filter height.

6.11 Cable Connection:

Over heating of cable is caused due to excessive current on account of loose thimbles or undersize cables. Cable must be designed for full load current and voltage drop on account of excessive length of cable used from generator to main control, though we try to place the generator near to main control though a change over switch. Thimbles of proper size are used as per value of load current or size of cable conductor. For design of cable, full load current is by taking power factor as 0.8

$$\text{So } I_{FL} = \text{Full load}/ 3\,V \cos\Phi$$
$$\text{For } 3\Phi = KVA / 1000\times3\times415$$
$$\text{For } 1\Phi = KVA/1000\times220$$

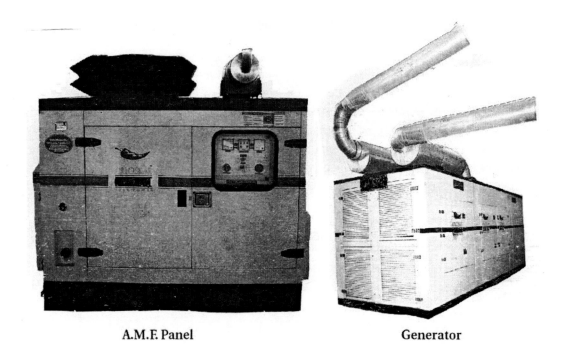

A.M.F. Panel   Generator

Size of cable is selected from correct rating table of copper cable and voltage drop as per chapter 5 of the book.

6.12 Change Over Switch:

We know that generator supply is required, when the supply from power corporation feeder fails to supply. So we need a device, which disconnect complete electrical installation which run from power supply and connect the installation with diesel generator supply. This device is called change over switch, which is just like a control switch ICDP or ICTPN. There are two type of change over switch namely on line or off line change over switch. A proper rated change over as per load current is used. On line change over switch automatically connect DG supply to load, disconnecting mains supply while off line change over switch break and make the mains and DG supply respectively manually.

6.13 Design of D.G. Set:

Total load in KW of the circuit of electrical installation in which D.G. set is to be installed to feed power supply must be evaluated. Taking power factor 0.8, KVA rating of D.G. set can be found by dividing KW by 0.8

For ambient condition i.e. temperature & humidity at altitude worsen the standard operating condition. The engine output is derated according to relevant standard. So a derating factor must be taken into account for selection of DG set for hilly area.

6.14 Battery:

Dry battery must be initially charged for 72 hours so that specific gravity of acid in side battery should be 1.23 to 1.28 Kg/liter. For long time interval during which generator remained idle i.e. not working, positive terminal of battery is disconnected and must be connected at the time of restart after long gap if battery is found charged at the time of start to run the self of D.G. set. We need 230V AC charger which is used after disconnecting both terminal of battery. Battery charger is provided with AMF panel to avoid unnecessary run of generator and to facilitate to stop and restart as per requirement. The battery charger is charged with dynamo/alternator fitted with engine.

6.15 Commissioning of D.G. Set:

After completion of DG installation on foundation together with exhaust piping fuel line fitting, cable connection to change over/ main control switch and proper earthing & panel fitting if provided out side, DG set is commissioned by service engineer of supplier on load. Diesel in tank lubricating oil in engine and charge of battery is checked as per commissioning check.

6.16 Operation:

Operation of DG set with both type of alternator is more or less same except presetting or checking of different component and assemblies e.g. checking of AVR in brushless and slip-ring and brush in slip-ring alternator. The prime mover i.e. diesel engine of suitable HP is required to give rated KVA of generator. Engine usually operates at 1500 rpm. After making pre check as per daily maintenance we start the DG set by applying 230V AC to exciter through control panel. Be fore transferring to load, we must perform voltage adjustment on all 3 phase to keep them equal by voltage hand regulator if provided. All 3 phase must have nearly equal and balanced load. One must keep only 5% unbalancing, though not more than 25% unbalancing is permissible.

So it is essential to check the rated current in each phase on transferring on load. It is essential to see the rated voltage in voltmeter of panel before transferring to load. As discussed earlier control panel is provided with HRC fuse. Proper ON/OFF line change over of proper rating is used. Panel is provided over load and short circuit protection device. Operation manual may followed strictly. Prior to put it on load.

6.17 Estimating & Costing:

To prepare the estimate for supply & installation of DG set, we design KVA rating of the DG set as per 6.13 of this chapter, specification and design of engine/alternator taking as per manufacturer specification available in market. Design of foundation, installation, earthing other accessories and equipments e.g. change over switch, cable etc, are done as per following manner.

6.17.1 Design of Installation: We have discussed in 6.7 of this chapter the installation is done at dry places and generator room as far as possible. The generator room with good ventilation is constructed, if not exist in building premises, after selection of site as per (6.6) of suitable dimension as per building specification and schedule of rate. The provision of construction of a generator room is made in the estimate.

6.17.2 Design of Foundation: As discussed in 6.7, the generator in kept on RCC foundation or MS I to ISMC channel. The length, width & thickness of RCC or length of channel is decided as per dimension of generator.

6.17.3 Design of Cable: The size of cable to connect DG set to change over switch & main supply control switch or main control switch of installation, to be fed, to change over is designed as per design of cable, as discussed in chapter O.H. line & cable of chapter 1 o in 6.11 of this chapter.

6.17.4 Design of Change Over Switch: After selection of on line or off line change over switch as per requirement, the capacity of change over switch is taken as per ampere load of generator as discussed in 6.12 of the chapter.

6.17.5 Design of Earthing: A suitable earth is selected as per 6.8 of this chapter and length of earth wire/strip is taken as per distance of earthing pit & place of DG set.

6.17.6 Detail of Measurement: After designing the detail of measurement is done for DG set, its foundation, length of cable, change over switch, earth wire/strip and summarize them with their brief specification.

6.17.7 Analysis of Rate: The analysis of rates for DG set, its foundation channel (if selected), cable, change over switch are made, if there exist no schedule of rates as per current market rates/manufacturer price list to know the item rates of each item to be provided.

6.17.8 Bill of Quantity: Knowing the rates and quantities of different items, the bill of quantity is prepared on standard format. The total of BOQ will give cost of installation of DG set.

6.17.9 Abstract of Cost: The abstract cost is also prepared as per standard manner, including expenditure, of miscellaneous item e.g. construction of room as LS with contingency for contingent item. The total of abstract of cost will give the cost of estimate or DPR etc.

7 | Lift

7.0 Introduction:

With reference to the need and invention of lift let us think about the creation of mankind & universe. In ancient days people used to live in group or kabila with concept of living together forming village in small bamboo or kacha houses. With the pace of development pacca residences, market, industries etc, came into existence Village converted into city / town & metropolitan city or kaval town. For the convenience and available of land people started constructing twin or three story buildings.

Further development and advancement of science and technology & scarcity of land resulted in construction of multistory, sky scraper or sky tower buildings in big town ship. In multistory building it was difficult to climb up or reach and carry or deploy goods and utilities on top floors, which resulted in invention of lift and elevator. The word lift means upward displacement or upward pulling while elevator means upward movement system. Now a days, modern lifts are used with advance protection against safely risk and more comfort or facilities to carry passenger their goods or bed ridden persons, old or handicapped persons from one floor to other in multistoried buildings and hospitals.

7.1 Definition:

7.1.1 Bottom Clearance: The space which the car-floor can travel below the level of the bottom lift-landing until the full weight of the loaded lift-car rests on the buffers.

7.1.2 Bottom Over Travel: The distance provided for the car floor to travel below the level of the bottom lift landing when the lift car is stopped by the normal terminal stopping device.

7.1.3 Buffer: A device designed to absorb or reduce the impact of the lift car or a counter weight at the external bottom limit of travel.

7.1.4 Car Leveling Device: An automatic device designed to cause the lift car to move at a reduced speed within a limited zone and to stop substantially at level with lift landing independently of varying loads.

7.1.5 Compensating Ropes or Chains: The ropes or chains suspended from the car frame or counter weight to counter balance the weight of the suspension ropes.

7.1.6 Controller: A device or group of devices comprising the principal components of control equipment.

7.1.7 Counter Weight: A weight or series of weight to counter balance the weight of the lift car or part of the load thereof.

7.1.8 Dual Control: A method of alternative automatic of car switch control, so arranged that either may be used but not at the same time.

7.1.9 Electro Mechanical Break: A break consisting of friction shoes applied to a brake drum by means of springs or weight and released electrically.

7.1.10 Emergency Stop Switch: A device designed to cut off power to the control circuit to cause the lift car to stop.

7.1.11 Final or Ultimate Limit Switch: An emergency stop switch designed to stop the lift car in the event of excessive over travel.

7.1.12 Flexible Guide Clamp Safely Gear: A safety gear in which the action on the guides is effected by means of rollers or cams applied gradually in an emergency.

7.1.13 Floor Selector: A mechanism which forms part of the control equipment in certain automatic lifts and is designed to operate controls which cause the lift car to stop at the required floor.

7.1.14 Gate Lock: A lock designed so that the door or gate, may only be opened when the lift car is in the landing zone or by a special key.

7.1.15 Gradual Wedge Clamp Safely Gear: A safely gear in which the action on the guides is affected by a screw and wedge or similar device applies gradually in an emergency.

7.1.16 Guide: The member used to guide the movement of the lift car or counter weight.

7.1.17 Guide Bracket: The part of a guide fixing which carries the guide seating or guide clips and bolts and serves to secure them to the buildings or structures.

7.1.18 Guide Shoes: An attachment to the car frame or counter weight for the purpose of guiding the lift car or counter weight.

7.1.19 Lift Pit: The space in the lift well below the level of the lowest lift landing served.

7.1.20 Lift Well: The unobstructed space within a enclosure provided for the vertical movement of a lift car and its counter weight including the lift pit and the space for top clearance.

7.1.21 Over Speed Governor: An automatic device which brings the lift car or counter weight to rest by operating the safely gear in the event of the over speed of lift car in a descending direction exceeding a pre determined limit.

7.1.22 Safety Gear: A mechanical device attached to the car frame or counter weight to stop and to hold the lift car or counter weight to the guide in the event of free fall or if governor is operated or over speed of lift car in the descending direction.

7.1.23 Top Over Travel: The distance provided for the car floor to travel above the level of the top lift landing when the lift car is stopped by the normal terminal stopping device.

7.1.24 Trailing Cable: A flexible cable providing electrical connection between the lift car and a fixed point or points.

7.1.25 Top Clearance: The space which the car floor to travel above the level of the top lift landing without any part of the lift car or its attachments coming into contact with any overhead steel work or other obstruction.

7.1.26 Travel: The distance between the bottom and top lift landings served.

7.2 Codes & Standard:

Design, specification, manufacture, installation, operation and maintenance are being regulated vide various standards in every developed countries. In India too various. IS codes have been enforced time to time which are enumerated as below.

(a) IS code 3534 1966 for outer dimension.

(b) IS code 1860 1968 for installation maintenance & operation

(c) IS code 5491- 1968 for installation & maintenance

(d) IS code 4666- 1968 for specification

(e) IS code 6303- 1971 for electric service

(f) IS code 6620- 1975 for installation operation & maintenance

(g) IS code 7759-1975 for door locking device

7.3 Rules & Act:

Manufacturing, installation & operation are controlled under law by some rules in some developed city & states and violation of these riles are governed by some act. Some of these rules and act are enumerated below.

1. Bombay lift act 1936
2. Bombay lift rule 1958
3. Delhi lift rules 1945
4. West Bengal Lift act 1955
5. West Bengal lift rule 1955
6. Rajasthan lift act
7. Rajasthan lift rules
8. Himachal lift act
9. Himachal lift rules

7.4 Classification:

Lifts can be classified in many ways according to drive, control system, use, speed, operation, lift lay out, shape & material of construction.

7.4.1 According to Type of Building: According to type of buildings, the lifts can be divided in following manner.

(a) For low & medium category residence.

(b) For offices, halls and upper category residences.

(c) For shops and departmental stores.

(d) For hospital, bed & passenger.

7.4.2 According To Drive: There are two type of drive employed. Accordingly, the lifts are classified as.

(a) Sheave drive lift.

(b) Drum drive lift.

7.4.3 According To Control & Speed: On this basis lifts are grouped as.

(a) Single speed AC control.

(b) Two speed AC control.

(c) Rheostatic control

(d) Variable voltage control.

(e) Thyrister control.

(f) Microprocessor control.

(g) AC servo drive control.

7.4.4 According to Floors & Speed: On the basis of floors & travel speed these lifts can be classified as.

(a) For four to five floors 0.5 to 0.75 m/s.

(b) For six to twelve floors to 0.75 to 1.5 m/s speed.

(c) For thirteen to twenty floors above 1.5 m/s speed.

(d) For above twenty floors high speed as required.

7.4.5 According to Type of Panel: Lift can be classified as per type of panel used as,

(a) All glass glazed panel (capsule).

(b) All steel panel

(c) Partly glass glazed.

7.4.6 According to Operation: Lifts can also be classified on the basis of their operation as.

(a) Car switch operation for big offices.

(b) Automatic push button operating for small offices.

(c) Selective and calculative operation with or without attendant for big offices.

(d) Signal operation incase of group of lifts with attendant for big office buildings

7.4.7 According to Use: Lifts are classified as per their use as,

(a) Passenger lift

(b) Goods lift

(c) Service lift

(d) Hospital lift

(e) Handicapped person lift

(f) Dumb lift

7.4.8 According to Dirive Point: Lifts can also be classified as power required as,

(a) Electric traction AC motor lift

(b) Electric traction DC motor lift

(c) Hydraulic lift

7.4.9 According to Shape: It can be classified as per shape of car

(a) Rectangular.

(b) Square

(c) Pentagonal

(d) Hexagonal

(e) Octagonal

7.4.10 With or With out Machine Room: Lift may be of two type.

(a) With machine room.

(b) Without machine room (MRL).

7.5 Lift Layout:

The place or layout of lift in building is very important and requires special attention for easy, straight and simple reach and centrally placing and the lay out is classified as below.

(i) Straight line arrangement up to three lift.

(ii) Alcob arrangement for more than three.

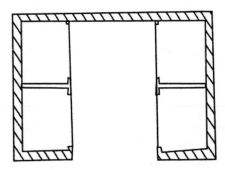

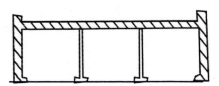

Three Lifts
Fig. 1 (A) Straight Line
Arrangement for Three Lifts

Four Lifts
Fig. 1 (B) AL Cove Arrangement
for more than Three Lifts

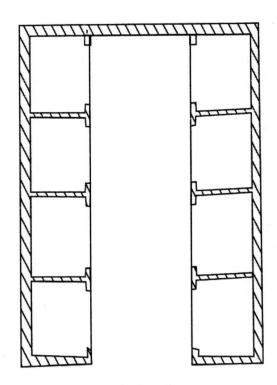

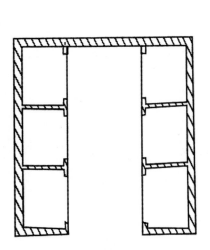

Six Lifts
Fig. 1 (C) Arrangement
for Six Lifts

Eight Lifts
Fig. 1 (D) Arrangement
for Eight Lifts

(ALCOB)

7.6 Components of Lift

7.6.1 Lift Car: The height of lift car should be 2m at least. It has two unit (a) Sling (b) Car Side frames of car is built up of steel angles plates and vertical diagonal stays duly riveted and suspension cross head is fixed at the top of frame, which is bolted with a housing for spring loaded top guide shoes. Sling is made of rigid bolted or welded rolled steel angle or channel section.

Lift well doors are usually made of SS steel or MS powder coated sheet of suitable size. The body of car is made with SS panel or power coated. MS enclosure while floor is made of steal covered with 2mm thick MS plate sheet. Milky white acrylic drop out false ceiling is constructed. Arrangement of CFL light, cabin fan & exhaust fan is made in ceiling for proper light & ventilation of car. The car has engraved specification plate, indicating rated load & passenger capacity. Some important instruction is also written on one panel of car.

Electrical connection to the car is made by a 10 core (multi core) hanging (trailing or traveling) flexible cable of length equal to half car travel plus 45m.

7.6.2 Lift Car Drive: Car drive may be either sheave or drum drive. In sheave drive one set of rope is used, which passes from car round a cast iron/steel V shaped grooved sheave (pulley) to the counter weight. In drum drive two set of ropes are used. One end of ropes are fastened by clamps on side of drum, while other end ropes are fastened to car and counter weight respectively. One set of ropes is wound clockwise for wrapping while other set of ropes anticlockwise for unwrapping. Sheave or drum is driven by a motor.

7.6.3 Advantage of Sheave Drive:

(a) If either car or counter weight comes in contact with buffers the drive ceases to avoid danger of hit of car with upper structure.

(b) It is cheap, simple and standard equipment used.

(c) It has large rope life and minimum level of noise and have greater degrees of silence.

7.6.4 Car Brakes: The most common form of lift brake is spring operated electrically released by a electromagnet. It is provided between motor and gear box i.e. on high speed side. The brake shoes are in two part and lined with Ferodo or similar material so that in the event of failure of one half, the other half is available for braking. The brake adjustment is done by adjusting spring tension and air gap between shoes and drum. D.C brakes are used having following of advantages over A.C brakes.

(a) Its operation is made smooth by connecting a resistance in parallel with coil of electromagnet.

(b) Solid magnetic cores are used in D.C brake while laminated cores are used in A.C brakes to minimize current losses.

(c) D.C brake is silent than A.C brake.

(d) D.C magnet pull is proportional to the square of gap while in A.C magnet it is constant.

7.6.5 Lift Gears: Because of thrust adjustment, the manufacturing problem low speed motor is practicable for 2m/sec. Minimum speed of motor is 1000 rpm, so gears rations, are used to get and low speed is 10 to 1. Cast iron box consist of steel worm and phosphor bronze, wheel totally enclosed in oil. The temperature of oil is not to increase 94°C with proper design of gear. Suitable bearings are used on sheave drum and gear shaft. Over speed governor is used with gears to cut power supply and apply brake, as lift gears are highly efficient and efficient lubrication are used in gear box.

7.6.6 Counterweight: It consist of sections of cast iron and secured by two steel tie rods with lock nut or split pins at ends. This counterweight balances the weight of the car and provides traction and helps is reducing size of motor with certain measure of safety, when landing on buffers. The hoisting ropes are secured to the counter weight frame by screwed eye bolts.

7.6.7 Guides & Shoes: Two guides are provided for car & two guides for counter weight to guide them to travel in vertical direction 3", 23/ 8" or 2" round steel guides are used with safety gear which when in operation huge loads are imparted on guides. In RCC structure the guides are fitted by plates and bolts, while in bricked wall expending bolts or eye bolts pass through brick work and are fastened by steel plates on the other side. Since the guides/bolts are exposed to dust and dirt from well and landings, the guides are lubricated to minimize wear of guides & shoes to reduce friction for comfortable riding. Dry guides with carbon lined shoes may also be used to clean the guides.

7.6.8 Rope & Roping System: The lift ropes are stranded of acid quality high carbon content steel wires with great strength and flexibility. The fiber cores of all lift ropes consist of jute hemp or manila impregnated with lubricants. The diameter of ropes varies between 6.50mm to 22mm with good factor of safety.

Several roping system are used to transmit power from winding machine to the car depending on the local condition, situation of winding machine, speed and load of car etc following two are common.

(a) Single wrap or one to one system:- The ropes wrap the friction sheave once 180° angle only.

(b) Double wrap or two to one system:- The ropes are securely anchored to one of the structural beams and then passes round a multiplying pulley fixed to counter weight pulley sheave a second multiplying pulley fixed to the car frame and finally to second anchorage. Load on sheave and bearings the peripheral speed of sheave is doubled. The speed of car, the traction is increased due to 360° wrap.

Ropes are fastened with car and counterweight by either spliced loops method or bulldog clips method or tapered babied sprocket method. Rope tension in each rope is equalized.

7.6.9 Lift Car Gates: Most common gates for lift car and landing entrance are the collapsible steel gates of over hang type. The gates are supported by ball bearing rollers, running on an overhead track, the pickets are guided by a self cleaning channel shaped bottom track.. Most of gates are single side opening type.

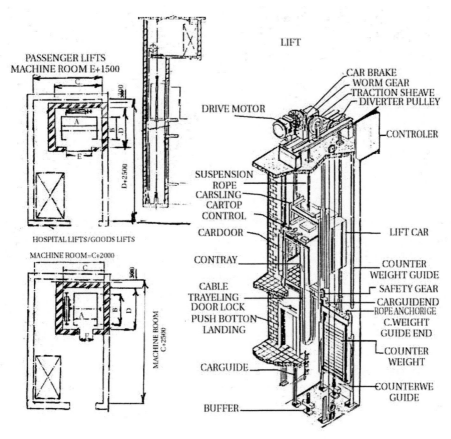

PASSENGER LIFTS
MACHINE ROOM E+1500

HOSPITAL LIFTS/GOODS LIFTS

MACHINE ROOM=C+2000

LIFT

- CAR BRAKE
- WORM GEAR
- TRACTION SHEAVE
- DIVERTER PULLEY

DRIVE MOTOR

CONTROLER

SUSPENSION
ROPE
CARSLING
CARTOP
CONTROL
CARDOOR

LIFT CAR

CONTRAY

COUNTER
WEIGHT GUIDE

CABLE
TRAYELING
DOOR LOCK
PUSH BOTTON
LANDING

SAFETY GEAR
CARGUIDEND
ROPE ANCHORIGE
C.WEIGHT
GUIDE END

CARGUIDE

COUNTER
WEIGHT

BUFFER

COUNTERWE
GUIDE

IS 3574 : 1976

HOSPITAL LIFTS

LOAD		CAR INSIDE		LIFT WELL		ENTRANCE
PERSON	KG	A	B	C	D	E
15	1020	950	2400	1700	3000	800
20	1360	1300	2400	2200	3000	1200
26	1768	1600	2400	2350	3000	1200

PASSRNGERS LIFTS

LOAD		CAR INSIDE		LIFT WELL		ENTRANCE
PERSON	KG	A	B	C	D	E
4	272	1100	700	1900	1000	800
6	408	1100	1000	1900	1600	800
8	544	1000	1100	1900	1900	800
10	680	1350	1100	1900	2100	800

7.6.9 Lift Doors: Now-a- days steel doors are used for landing protection instead of gates as they are quicker in operation, noiseless, and have improved appearance of lift and eliminate draught. There are following type of lift doors.

7.6.9.1 Hinge Doors: They are used on landings only and they may be single door or double center opening door type but opening is equal to lift car width.

7.6.9.2 Sliding Doors: These may be single panel or double panel with vision panel and safety glass. Single panel doors are used where opening are about 3 feet (910mm). There are two type of double panel doors. In one type both doors slides to one side only and one slides behind other. So the speed of former is twice the speed of other so they come at fully opened position.

In other centre opening type one panel slide to left other to the right. The clear opening is half the car width. Rubber bumpers are fitted on the edge of each panel to minimize impact during closing. The opening time is half that of other type. These panel doors are very popular for passenger lift cars where high speed is required.

7.6.9.3 Collapsible Steel Shutter Doors: These are sliding doors with large number of narrow panel which collapses into a small space. It gives wide opening with no draught and it is fire resistant, as it is made of steel or aluminum alloy. Door, gates may be either manually opened or opened with power A/C or DC electric motor drive. Both motors move the car doors through vee belt or chain drive.

7.6.10 Lift Door Locks: The most common form of lock is mechanical lock with electrical inter lock. So the gate / doors can not be opened from the landing side until car is level with landing and car should not move unless all gates/doors or closed & locked.

7.6.11 Lift Motors: Lift motors running on 1000 to 500 rpm must be quite and of lower kinetic energy to sustain rapid acceleration & retardation and starting torque must be equal to twice full load torque. So single speed motor is suitable for low speed up to 100 ft/m with starting KVA should not exceed 5KVA/HP. A small flywheel for adjusting inertia is attached for smooth operation. Better speed regulation, lower starting current and smooth acceleration is essential.

Two speed pole changing squirrel cage motor is used for car speed between 100ft/m to 200ft/m. Two separate stator windings in the same slots are used to give that pole change. The high & low speed winding are cut in and out of circuit by means of contactor.

7.6.12 Lift Controllers: Every motor has certain inherent characteristic. A controller is used and so designed to take care of the functions not incorporating the motor design. The various function not incorporated are.

(a) To limit the current during acceleration of the motor.

(b) To limit torque (current field strength) during acceleration.

(c) To change the direction of rotation.

(d) To regulate the speed of rotation.

(e) To limit the load of motor.

(f) To start and stop motor at fixed point.

(g) To disconnect the motor on failure of supply.

(h) To stop the motor.

Controllers are designed by manufacturer for following additional control for lift panel.

(a) The lift should not operate when any landing or car doors is open.

(b) To guard a possibility of the car or controller weight to encounter any object.

(c) To guard against damages to AC motor due to low voltage phase reversal and phase failure.

(d) To safe guard the motor against possible damages caused during operation of reversal of contact.

7.6.13 Magnetic Contactor: The Copper contactor faced with silver tungsten or silver cadmium are used. The surface of contactor must be clean and free fire from oxide scale. Making and breaking of contactor or results into weld or freeze of contact. Hence lift duty controller uses copper carbon contacts combination.

7.7 Wiring:

Suitable wiring is made in machine room, landing & car for proper installation and proper function for all equipment of the lift. Wiring is also done for providing for light and air circulation fan, exhaust fan etc. in lift car and machine room.

7.8 Earthing:

Separate earthing of lift and main control panel or switch board is done as per earthing chapter of volume-1 of the book. All controls, motor, panel board etc. must be securely earthed as per specification.

7.9 Safety Feature:

Many safety measures are taken with various components of lift, we will discus them in brief as below.

7.9.1 For Car & Counter Weight: Safety gear used in every lift is operated by an over speed governor to open break circuit of lift car. It may be installed in following 3 ways.

(a) Two cam and two cam shaft system:- Safety gear uses two cams with two camshaft which are linked together and rotated in opposite direction, when safety rope is pulled which brings cam in contact with guides through the speed governor, during over speed in down direction in case of breaking or stretching lifting ropes.

(b) Gradual wedge clamp system:- In case of lift car speed 200ft/m (1m/sec) gradual wedge clamp system is used. Governor rope passes over the governor sheave is driven by car rope carrier fitted with car frame. In case of over speed of car the governor jaws pulls the governor rope and clamps exerts retarding force upon guide to stop car.

(c) Flexible grid clamp system:- In this system spring inertia is incorporated in the linkage system to prevent impulses on governor rope.

In instantaneous cam type system, the stopping time is small, but in other two system stopping time is less.

7.9.2 For Terminal Stopping Switches: Every electric lift must be provided with an upper and lower terminal stopping switch to stop the car automatically with in top and bottom over travels from the contract speed. These switches are fitted to the high speed car operated by ramp in well and low speed car they are fitted in well and operated by a car ramp.

7.9.3 Buffers:- Two buffers are fitted in the lift well under the lift car, the center of ground of car and one or two buffers are fitted under counter weight. In the event of failure of final limit switch the car counter weight strike buffer.

For speed up to 1m/sec buffers may consist of helical springs or volute spring size capable of stopping the loaded car without permanent distortion.

For speed above 1m/sec a spring return oil buffers are used with piston and cylinder. Cylinder has number of holes in the wall. The number and size of hole downward are less compared to upper area. Spring is connected with piston rod in upper part of piston. After striking car with buffers the spring is compressed which acts as cushion for rod and piston. As the piston moves downward it pressers oil beneath, which comes out through holes and flows to outer casing and offers little resistance to the car in upper chamber of cylinder, so it accelerates when piston closes last hole the car stops. Piston is forced upward when car moves away by spring to bring normal run of car. Gravity return oil buffers are also used below counter weight to stop the car from striking upper structure of well.

7.9.4 Floor Leveling: Accurate leveling of car at various landing is required for good lift service. In automatic lift leveling is done by settings of slowing & stopping device and the mechanical brake as per variable load of occupants in car. The required speed of car is obtained with an ordinary shunt DC motor.

7.9.5 Directions Switches: Two way switch is necessary for each intermediate landing for both up & down direction, but a single way switch is required at each end terming landing to be operated by ramp in car. Top position of two way switch works in down ward journey & bottom of contactor for up ward journey. While, when lift is stationery, the direction switches remains at center or off position.

7.9.6 Floor Selector: Floor selector are installed in machine or motor room are of different type. In one case the bottom portion in car position indicator, selector arm rotate in with lift car movement in well.

For speed of 300 ft/min and above it is mechanically driven from the car by a toothed steel tape. The selector also controls the car and landing signals and indicators and opening of the car doors. Various methods are adopted for securing accurate final leveling by operation of floor selectors & stopping switches.

7.9.7 Car Indicator: Indication of car's motion at various landing is provided for waiting passengers at landing. Indicator shows the direction of travel and actual position of lift car in well during car movement.

7.9.8 Car Control System: There are number of systems of car control, depending on type of service required, type of building and car speed. System must be such that the car must be in complete control of passengers in car or on the landing. There are following control systems.

(A) Automatic control: A single cell/button is fitted at each landing and a set of buttons comprises one for each floor and other provided in car. The car is controlled by passengers in car.

(B) Semi automatic control: In this system up, down and stop buttons are fitted in car and at each intermediate landing. This system is used for two floors only.

(C) Automatic collective control: In this fully automatic system up and down buttons are fitted at each intermediate landing, single button at terminal landings and full set of buttons together with emergency stop buttons fitted in car. Up & down calls are answered during up and down journey in sequence of pressing of buttons, so it is called direction collective control system.

(D) Duplex collective control: This is used, where two lifts are installed in adjacent well. A common landing button serves for both lift but the call made by passenger on landing is accepted by the lift car which is nearer to that call traveling in that direction. Of course coming signal is illuminated at landing.

(E) Triplex collective control: This is used where three adjacent lift are installed with a common landing call button system in similar manner as in duplex collective control.

(F) Dual control: The lift may be worked either by an attendant operating the car switch during peak load or passengers operating buttons during light traffic, depending upon the position of transfer switch, which can change made of operation. During operation by attendant in peak hours, service is improved as attendant of car act according to need.

(G) Dual collective control: This is a collective control with additional facilities of being operated by an attendant during busy hours or on special occasion. The closing of doors and the starting of lift, switch and buttons are controlled by attendant in the car. The calls from landing are dealt automatically and in order as in collective control, but the attendant may by-pass the calls, if car is full, by a bypass switch or pushed button. A transfer/change over switch converts to collective control as the attendant leaves the car. If no car is registered, car will return to home lading i.e. ground floor.

(H) Signal collective control: This combines the feature of signal and collective control and therefore may use with or without an attendant. It can be used with one, two or three cars, .normally as signal control is used. at night by passengers without attendant, i.e calls of passengers at landing by pushing up or down buttons, keep the button illuminated until

call is answered. On collective control the car answers the nearest call from car or landing in the direction car travel. All calls are answered in rotation & sequence. When car reaches highest floor the direction is automatically reversed.

7.9.9 Selection of Drive: As discussed, the drive is either D.C. motor drive or A.C. motor drive. The DC drive is obtained by separate DC excited motor. Excitation DC voltage is obtained either motor generator or rotator converter in Ward & Leonard system or W/L system or by thyristor rectifier (3 phase bridge connection). The thyristor converter has following advantages over W/L system.

(a) Extremely good dynamic control response i.e. better regulation.

(b) High over all efficiency with negligible no load losses.

(c) Very compact so less in weight.

(d) Negligible maintenance.

(e) Less down time as it has quick break down maintenance.

(f) Small noise level with no vibration generation.

(g) Low initial cost.

7.9.9.1 AC Drive System: In this system squirrel cage motor is used and it requires less maintenance and it is cheap in cost as compared to DC motor, but its needs regular maintenance. Now-a-days, with use of semiconductor thyristor and microprocessor layer (electronic). AC drive system is more economic and stable. Following two type of AC drive system are used.

(A) Three phase thyristor & electronically controlled non synchronous induction motor with voltage, phase or tachometer controlled or using 3 phase thyristor voltage regulator controlled.

(B) Three phase squirrel cage synchronous induction motor with variable voltage and variable frequency system control.

7.10 Inverter or Converter:

Inverter is fundamentally convert fixed frequency AC supply into a variable frequency AC supply for above B type AC drive. These are voltage source or current source inverter depending upon the voltage or current parameter in DC link of the inverter. Voltage source inverter can be used for a single motor. The out put voltage may have quasi square or pulse width modulation (PWM) waveform for keeping voltage/ frequency ratio constant.

Cyclo converter are also used to produce variable frequency from direct AC supply without any intermediate DC link and used for low speed.

7.11 Programbeable Logic Control(PLC):

A large number of relays, contractor, timer etc. are required to perform staking control each other interlocking control function to avoid motor wear & tear due to heavy switching duty and for increased frequency of operations a programmable logic control (PGL) system is used to give following advantages.

(a) High flexibility.

(b) Easy erection.

(c) Less maintenance as only static components are used.

(d) Test and need of cabling & inter connection is greatly, reduced.

(e) Easy check on fault.

7.12 Test:

Following test are for performed by the installation engineer in presence of inspector of lift, prior to put the lift into service for normal and regular operation and testing are also done at subsequent inspection.

(i) **Machinery & Safety Gear Test:** Machinery & safety gear operate satisfactorily with in limit with full load in lift car.

(ii) **Wiring & Connection Test:** The electrical wiring and connection is tested for earthing, insulation and general soundness. Keeping brake, limit switches, buffers, safety gear, drive gears and speed governor etc. in operation.

(iii) **Traction Drive Test:** Lift is ascertained by a trial descend with 15 time in the full load, whether the friction between the ropes & sheave is sufficient.

(iv) **Runway Test:** This test is performed with all electrical apparatus operative, except for the over speed contactor or cut out on the governor. For lift operating directly on AC, the governor will be tripped by hand at maximum speed.

(v) **Over Test Voltage:** Insulation of electrical parts of all operating and similar device shall be tested to withstand test voltage 10 time the working voltage or maximum 2000V for one minute.

(vi) **Subsequent Safety Gear Test:** On each subsequent inspection on the safety gears, gears shall be tested with lift car stationary lowering lift car for satisfactory working of safety gears.

7.13 Installation of Lift:

Lift well and machine room are prepared with proper lift pit, opening of landing, various components of lift such as motor drive, drum, sheave, control panel etc. and then guides for lift car & counter weight are fixed on the wall of lift well. Ropes are hanged over sheave. Then lift car & counter weight are suspended after proper assembly of frame panels, guide shaft etc. After proper fixing of lift car & counter weight the doors or gates are erected in car & at opening of landing. Finally after testing the derive, and control, finishing of lift car is done to give required shape of lift car i.e. painting, if required on MS panel and sections etc. fixing of various switch, puss button, indicators, required plates, fixing of light, fresh air fan and other devices as sound system, ARD system etc. The electrical wiring is done for making proper connection.

7.14 Required Civil Construction For Installation:

For installation of lift the construction of following is made with building.

(a) **Lift Pit:** Lowest portion of lift well is prepared below ground level landing. Brick blast concreting is done as required by manufacturer of lift.

(b) **Lift Well:** A hollow rectangular vertical passage well or shaft is provided at required place for installation of lift. One lift well each for one lift is made in straight line or alcob layout for multi lift installation.

(c) **Over Head Clearance:** On the last floor of lift, the height of well should be kept more than the height of the floor to give over head clearance to accommodate lift car, while stopped at top floor.

(d) **Floor Landing:** At every floor an opening is made in lift well. Of course it is closed by doors or gates.

(e) **Machine Room:** An room is made either above at top floor or adjacent to the well at ground floor as per type of lift is used to accommodate machines, panel, various components, sheave & drums. Machine room is not required for MRL lift.

(f) **Lift Lobby:** A lobby is required to be constructed at each floor for waiting passenger.

(g) **Over Head Beam:** At the top of lift well below its roof an over head beam of required strength of proper reinforcement is made for suspension of lift car to accommodate sheave.

(h) **Opening In Lift Well's Roof:** Two or one opening of sufficient size is cut or made in slab of the roof of lift well for running of ropes.

7.15 Selection of Lift:

As we have discussed various type of lifts in 7.4 of this chapter, we select a proper type of lift as per our requirement i.e use, drive, control, speed and floor. In a hotel, shopping complex (mall) etc. (glass glazed) capsule type car is used otherwise steel car lift is used, which is useful with security point of view. Hydraulic lift were used, where electric supply is not available. These were useful for two or three floor. After designing the number of lifts, their proper layout arrangement (straight line or alcob) is selected.

7.16 Breif Specification:

The specifications of lift consist of the type of lift, capacity, rated speed travel, number of floor surved inside, size of lift well, lift pit depth, head room, clear inside size of lift of car dimension & position of machine room, position of counter weight, type of control, type of operation, type of controller, size & material of car, entrance doors or landing doors, constructing design & finish work of car body, type of signal system, size and position of landing entrance, type of floor, direction, lift in order or out of order display/indicator, type of power supply. Specifications must also describe availing of additional facilities, if required e.g. Automatic Rescue device (ARD), Emergency power supply (D.G. set), intercom system, musical or tele voice sound system, car chime etc. Environment condition & neighborhood temperature may also be specified.

7.17 Additional Operation:

The lift may have following additional operation facilities on demand and are provided as optional if required.

7.17.1 Anti Nuisance: If controller detect two many vacancy in lift car this device cancels all floors calls received on pressing call buttons on landing by passenger at landing and by passes floor landing.

7.17.2 Independent Service: A key switch is provided which, if turned on, refuses lift to respond any landing call except car calls.

7.17.3 Automatic Resceue Device (ARD): This is battery operated device which operates in case of power failure, with in few seconds, subject all safeties being in position and car is moved at slow speed to the nearest landing and the doors open.

7.17.4 Tele Voice System: This is a sound system, which announce the numbers of floors on which car is about to reach automatically before the opening and about next floor after closing of door after stopping at any floor.

7.17.5 Musical/PA Sound System: Arrangement for musical, PA sound system is made with speakers in car.

7.17.6 Fire Man Service: Upon switching on two fire man switch on the landing of pre determined floor, the lift car rushes to that floor for passenger evacuation or fire fighting in case of fire accident.

7.17.7 Fire Alarm Home Landing: This device when actuated by fire accident device installed in building & the lift car rushes to pre determined floor for emergency purpose of evacuation or fire fighting.

7.17.8 Car Chime: This device actuate playing of car chime twice moving in down direction and once for once in up director to help the vocally handicapped person.

7.18 Design of Lift:

To design the number and capacity of lift, a detailed study of building and occupant i.e. use of lift the following technology analysis or design are done on basis of standard norms.

(a) Occupancy: The population is calculated on the basis that one person needs 9.5 m².

$$\text{Population} = \frac{X}{9.5}$$

Where X = carpet area of building in m².

The suitability or capacity (passenger) lift is decided as per handling capacity of lift. Greater the handling capacity better is lift arrangement. The handling capacity (H) for non residential building and residential building are taken as about 7.5 and 15 to 25 respectively. The handling capacity is found by.

$$H = \frac{60 \times 5 \times Q \times 100}{T \times P}$$

Where   Q = Passenger capacity of lift

T = Waiting time

P = Population (occupancy)

We assume return traveling time 5 minutes i.e 5 x 60 seconds time for calculation.

So Q is calculated for 5 minutes for best handling capacity of residential or non residential building. Further choosing 6 or 8 passenger capacity of lift number of lift required can be found. Knowing quantity of lift the straight line or alcob layout arrangement is selected lift waiting time (T) for passenger must be minimum at different floor.

(b) **Quality of service:** The quality of service will be better, if lift waiting time (T) is minimum and the number of the waiting passengers shall be minimum at floors. The following classified standard of quality of service for lift is adopted.

Waiting Time T (Seconds)	Quality of Service
20 s to 25 s	Very Good
25 s to 35 s	Good
35 s to 40 s	Average
40 s to 45 s	Bad
More than 45 s	Unsatisfactory

Waiting time (T) is found by.

$$T = \frac{RTT}{N}$$

Where   RTT = Return travel time (time to come back starting from any floor).

And   N = Number. of lifts

Keeping N constant, if RTT is reduced, then T and quality of service can be improved. Handling capacity can be increased, if waiting time of the lift at floor T is reduced. By adjusting speed of lift and stooping time of lift at floor return traveling time RTT can be controlled to give better quality.

Example: In a fifteen storied building with plinth and carpet area of ground floor is 1100 m² and 950 m² respectively. Design stable lift with specification as below.

Lift capacity = 20 persons

Speed of lift = 2.5 m / s

RTT = 165 s / trip

Solution:- Population P $= \dfrac{\text{Carpet area of floor x No. of floor}}{9.5} = \dfrac{950 \times 14}{9.5}$

$= 1400$

Average lift capacity $= 80$ % of lift capacity

$= 80$ % x 20

$= 16$

The suitability for capacity of lift is decided as per handling capacity H

$$H = \dfrac{60 \times 5 \times Q \times 100}{T \times P}$$

(a) If number of lift is taken as 4

$$T = \dfrac{RTT}{N} = \dfrac{165}{4} = 41 \text{ second}$$

Waiting time T 41 sec come in bad category

$$H = \dfrac{60 \times 5 \times 16 \times 100}{41 \times 1400} = 8\%$$

This is suitable for residential building not for non residential building.

(b)   If number of lift as taken as 6

Then   T = 165 / 6 = 27.5 s which is good waiting time

So $H = \dfrac{60 \times 5 \times 167 \times 100}{27.5 \times 1420} = 12\%$

Though handling capacity is not completely satisfactory, but keeping N fixed the lift of higher RTT is selected.

7.19 Estimating & Costing:

Following steps are adopted for proper estimating & costing.

7.19.1 Design of Lift: Numbers of lifts are designed as per type & population of building by assuming 6 to 8 passenger capacity as per 7.18 of this chapter.

7.19.2 Selection of Lift: As per type of building & service required a proper lift with thir specification is selected as per prevailing condition as discussed in 7.15 of this chapter.

7.19.3 Detail of Measurment: after finding the number of lift required, it is summarized with brief specification under detail of measurement, The capacity of emergency DG set, change over switch are also taken in detail of measurement after proper design as per chapter VI of this book., Control switch of lift and electrification required in machine room, lift well and generator room and earthing etc. are also designed and estimated as per volume I of this book and summarized in the detail of measurement. Thus the quantities of each item are found by summing them.

7.19.4 Analysis of Rate: A proper analysis of rates for lift is prepared after obtaining rates of manufacture of lift, if their exist no schedule of rates

7.19.5 Bill of Quantities: A proper bill of quantity on standard format with complete specification of lift is prepared to find the cost of lift.. A separate bill of quantity is also prepared for earthig, electrification of machine room, lift well and DG room etc including the provision of D.G. set & its accessories if required..

7.19.6 Abstract of Cost: Finally the abstract of cost is prepared including BOQ of lift, BOQ of electrification of lift well, machine room and D.G. room& BOQ of D.G. set. The provision of contingencies, and centage charges are also included together with lump sum provision of any unforeseen items The total of this abstract of cost shall give the complete cost of Installation.

8 | Computer Networking

8.0 Introduction:

A computer networking is done to connect two or more computers, to allow them to communicate with each other and to share their resources and information. That is if one computer have Printer, scanner, CD writer, a back up system, audio / video CD or DVD player connected to the computer, one can print, scan, burn CDs, make or listen audio voice or song, see video clips or films at other computer, without their resources, if connected with first through network. Similarly we can access files data, information, available at one computer by other computers of network.

On the basis of tele type machine, which was developed in 1940, George stibitz. APRA designed in 1969 APRANET (Advanced Research Project Agency network), for US department of defense.

As computer is very common tool, now-a-days, as it has wider advantage and application in most of important buildings, one requires computer networking installation. PAN, LAN are very common in buildings or group of buildings occupied by various institution and enterprises. In a big campus having large number or group of buildings, CAN is needed. So as an engineer associated with construction, maintenance of building one must know about design, installation & maintenance of computer networking. Use of internet work or internet by people has become essential and useful for the scope of video conferencing between various employees of any institution of government and private sector.

8.1 Classification:

Computer networking can be classified in many ways.

8.1.1 Connection Method: According to this method hardware & software technology used to interconnect individual device in the network. Connection method uses either wired or wireless technology. The wire may be twisted, pair wire, coaxial cable or fiber optics. Examples of wired technology are Ethernet, LAN, PAN, Home PAN and G. hn.

The wireless method uses radio wave or impaired signal microwave etc. Examples are Terrestrial microwave, communication system, cellular and PCS system, wireless LAN, Bluetooth and wireless web.

8.1.2 Scale Method: Computer networking often also classified on the basis of scale of area in which networking is done it may be classified as below.

8.1.2.1 Personal Area Network (PAN): It is a network used for communication among computer devices such as PC, CPU, UPS printer, scanner, fax machine, telephone, PDAS etc.

8.1.2.2 Local Area Network (LAN): It covers small area e.g. home, office or small group of building (as school, airport) the number of computers in a buildings or small group of buildings are interconnected through LAN.

8.1.2.3 Campus Area Network (CAN): It is interconnection of LANs with in limited area. CAN is larger than LAN but smaller than WAN e.g. university campus based CAN connecting various campus building having LAN.

8.1.2.4 Metropolitan Area Network (MAN): It connects two or more LANs or CANs, but limited in the boundary of a town, city. Routers switches and hubs are provided in MAN and are interconnected.

8.1.2.5 Wide Area Network (WAN): This computer network covers a broad area with communication links across metropolitan, regional or national boundaries. This uses routers, modem with VSAT antenna and public communication links. The largest and a well known example is internet.

8.1.2.6 Global Area Network (GAN): The global area communication network, developed by several group e.g. mobile communication, across wireless LANs, satellite coverage area etc.

8.1.2.7 Virtual Private Network (VPN): It is a computer network in which some links between nodes are carried by open connection or virtual circuits in large network like internet instead by physical wires. VPN can be used to separate the traffic of different user communicates over an underline network with strong security feature.

8.1.2.8 Storage Area Network (SAN): A computer network communicates in certain storage area is called SAN.

8.1.3 Functional Relationship (Network Architecture): Computer network can be classified according to functional relationship, which exists among the elements of the network e.g. Active network client server and peer to peer (work group) architecture.

    8.1.3.1 Peer to Peer Networking: A network is said as peer to peer network, if most of computer are similar and run work station system. Each computer hold there files and resources and other computer can access other computer files, data, resources etc. if its computer is ON e.g. If a printer is connected to a computer A, and other computer wants to use that printer computer, A must be ON.

    8.1.3.2 Client / Server Networking: A computer network is called as client / server networking, if at least one of the computers referred as server, is used to serve other computers, referred to as clients. The different type of device may also be part of the network together with computer.

    8.1.3.3 Advantages of Client / Server Networking:

(a) Server holds files and resources and client can access them without caring weather other clients is ON. Server is always ON.

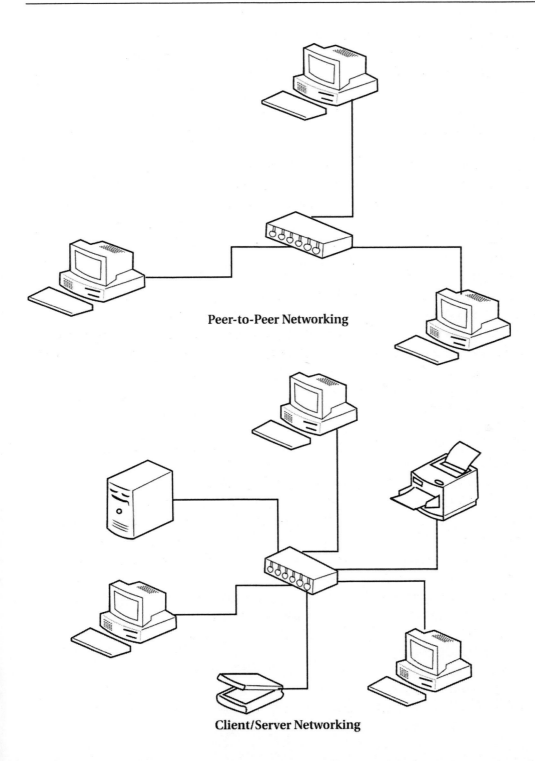

Peer-to-Peer Networking

Client/Server Networking

(b) Security can be created, managed and highly enforced. User is provided with some credentials including user name and a password.

(c) It has centralized backup, internet capability, internet monitoring etc.

8.1.4 Network Topology: Computer network may also be classified according to network topology, upon which the network is based e.g. bus network, star network, ring network, mesh network, star bus network, tree or hierarchical topology network.

8.1.5 Method of Data Used: A computer network is classified as digital or analogue.

8.1.6 Idustrial or Commercial Networking: The networking may be commercial or industrial. In the Industrial networking, the system is dust and oil free designed for high humidity and free from exposure to water, which are usually present in industrial area to safeguard network against them. The networking without these is commercial networking.

8.2 Internet Work:

Internet work is the connection between two or more distance computer networks or networks segments among or between public, private, commercial, industrial or government network via a common routing technology. It is often called internet. It operates at layer 3 of the OSI Basic Reference Model e.g. router. Inter connected network uses the internet protocol.

There are three type of internet work (a) Intranet (b) Extranet (c) Internet.

8.2.1 Intranet: Intranet is a set of network under sum single administrative entity or internal network of a organization. It uses (IP) internet protocol and IP based tools e.g. web browsers and file transfer applications. A large intranet has one web server to provide users with connected organization file.

8.2.2 Extranet: Extranet is a internetworking which has limited connection to the network of one or more trusted organization in addition to limited scope a single organization. Extranet may also be categorized as CAN, VAN, WAN or other type network i.e. single LAN must have connected with extranet network at least.

8.2.3 Internet: It consists of worldwide inter connection of government, academic, public and private networks based on the networking technology of internet protocol suite. Internet is written as prominent with capital letter 'I'. Its service provider and large enterprises exchange information through Border Gateway Protocol (BGP).

8.3 Basic Hardware Component or Computer Networking Device:

All network uses basic hardware components (building blocks) to inter connect network nodes, their wires (cat cable, optical fiber cable) or wireless through microwave links. Ethernet card is required for connection. The various components are NIC, bridges, hubs, switches and routers etc.

8.3.1 Network Interface Card (NIC): It is a piece of computer hardware used to allow computer to communicate over a computer network. It often provides a low level addressing system using MAC addresses.

8.3.2 Network Repeater: It is a electronic device which amplify signal i.e. it receives signal and retransmits it at a higher power level to other side of obstruction, so that the signal can cover longer distances without degradation, usually used for cable of more than 100m length. It works on OSI layer -1.

8.3.3 Network Hubs: It contains multipurpose port when a packet reaches at one port, it is copied unmodified without change of destination address in the frame, to all ports of the hub for transmission. It connects multiple ethernet segment to act as single segment. Every attached device share same address.

8.3.4 Network Bridge: It connects multiple network segment at the data link layer (or layers) of model. It do not copy traffic to all ports as hubs, but first find MAC address reachable through specific ports than send the traffic for that address only to that port. Bridges do send broad casts to all ports except the broadcast receiver port.

8.3.5 Network Switches: It is device that forward & filter chunk of data communication (OSI layer2 data) between ports based on MAC addresses in the packets. It is not capable of routing traffic based on IP address (OSI layer 3) as router. A switch has numerous ports, to connect most or all networks. A switch is to encompass the routers, bridges and devices to distribute traffic on loads (e.g. web URL identifier). It may operate at one or more OSI model layers including physical, data link, network transport (end to end). A switch that operate simultaneously at more than one of these layers are called multilayer switch.

8.3.6 Routers: It is a networking device that forward packets between networks using information in protocol headers and forwards tables to determine the best next routers for each data packet. It works at net work layer (OSI layer 3) and the internet layer of TCP/ IP. Unlike gate way it can not interface different protocols.

8.3.7 Network Wireless Access Points (WAP): It is a device in computer networking that allows wireless communication devices to connect to a wireless network using Wi-Fi, Bluetooth or related standards. It is connected to a wired network and can relay data between wireless devices as (as computer or printer) and wired devices on network. WAP include a router or broad band modem and internet switch. A WAP has limitation of communication with about 30 clients with in radius of 100m. However it can be affected by variable factors like, its indoor or outdoor placement, its height above ground, near by electronic device operating on similar radio frequency, type of antenna, operating frequency and power output of device. However the range of WAP can be improved or amplified by repeaters and reflectors. WAP opened to public is referred as hot spot.

8.3.8 Industrial Wireless Acces Point: It is rugged with metal cover and mount and is kept free from exposure to water, dust and and it must sustain wider & higher range of temperature and high humidity .

8.3.9 Network Gateway: This device uses different protocol setup sitting at a network node for inter facing with another network. It works on OSI layer 4 to 7.

8.3.10 Network Multiplexer: This is a device which combines several electrical signals into a signal.

8.3.11 Modem: This is a device which modulates an analog carrier signal (e.g. sound) to encode digital information that also demodulates such carrier signal to decode the transmitted information, as a computer communicating another computer over the telephone network of server.

8.3.12 ISDN Terminal Adapter(TA): It is a specialized gateway for ISDN.

8.3.13 Line Driver: A device used to increase transmission distance by amplifying signal, base band networks only.

8.3.14 Digital Media Receiver: It is a device which connects a computer to a home theatre.

8.3.15 Protocal Convertor: It is device that converts between two type of transmission asynchronous and synchronous transmission.

8.3.16 Fire Wall: It is software or hardware put on network to prevent some communication for forbidden by network policy.

8.3.17 Proxy: It is computer network service which allows clients to make indirect network communication to other network.

8.4 Open System Interconnectiong Model:

Network is defined by standard of OSI reference for communication. It consists of following seven layers. Upper layer, (Application, presentation) is used for application while lower (data, link, physical) focus on signal flow of data from origin or designation. Various OSI layers are.

Layer-1 Physical network layer primary data deals with physical characteristic transmission of data.

Layer-2 Data link layer set standard for data being delivered across a link or menu.

Layer-3 Network layer deals end to end delivery of data packet.

Layer-4 Transport layer includes protocol to allow it to provide function error receiving segmentation of reassembly.

Layer-5 Session layer stabilizes for start control and end links or conversation.

Layer-6 Presentation layer focuses on defining data format e.g. text, page(Picture received in e-mail) gift and binary.

Layer-7 Application layer defines medium that communicate software and any application needed to communicate to other computer.

8.5 Installation of Computer Networking:

As a construction engineer, we must have sufficient skull of computer networking in a building. Knowing the requirement of user/occupant of building we select type of network.

8.6 Selection of Network:

With study of map (plan) of building and discussion with user of building, we can know the occupancy (the number & type of occupant), we find out the number and place of computer required to be installed i.e. The number of nodes and summarize them in a table. We also discuss whether some or all computer are to be connected with each other or individual computer (not interconnected) is required to decide that only PAN is required or both PAN & LAN are required. If there are number or group of building in a campus, then CAN is required.

8.7 Networking in Media Center:

As the media person uses computer and internet work, we provide computer networking at media centre. Every table is provided to a correspondent/media person of any channel or news paper etc. is considered as node for networking. Both power outlet and modular information outlet are provided at each table. Power outlet (i.e. A 16A socket with two or three 6A socket) is require for U.P.S., Printer etc. and a MIO for broad band/ Wi-Fi/ Wi-mac connection of service provider or of own VSAT modular converter provision. These are given through submain or internet working as discussed earlier.

8.8 Design of Computer Network:

8.8.1 Design of PAN: UPS, Fax, printer, scanner, printer cum scanner, printer cum Photostat, scanner and other device is to be connected with computer, PAN is needed. Also, if number of computers in a building, not to be interconnected and are to be operated independently, then also PAN is needed.

If the power to each computer is to be supplied by individual UPS, then be provide a power socket with two or three 6 A socket on a switch board which is fed through 2.5 sqm copper submain as discussed in chapter V of volume I of this book. The power socket is require for UPS and other socket are used for printer, scanner etc. For PAN some patch cod are needed as per pots on CPU to connect monitor, key board, mouse, speakers, web camera etc. and UPS to CPU. If it is necessary to connect the computer with internet for internet work i.e. to have internet capability on computer, we need a broad band or WI-FI connection (link) from service provider as BSNL or Reliance etc. or we can access internet through VSAT antenna and modem, router or Bluetooth device.

8.8.2 Design of LAN: If number of computer in a building are to be connected with each other to communicate data information, we need LAN. If there are some computers installed in a room or hall, then peer to peer networking can be used while if the computers are situated at distances we use client / server networking. The disadvantage of peer to peer networking apart from no security is that one can only access the other computer if it is ON as discussed 8.1.3.3.

8.8.3 Design of Server: A suitable serve selected and housed at a suitable central place, preferably near to video conference room, if provided. A UPS of suitable capacity is installed at server room and submain wiring is done in conduct with 2.5 sqm suitable copper wire preferably concealed conduit, from server room (UPS) to each nodes. Separate earthing is done for UPS and earth wire shall run with submain wiring to UPS outlet box (with 16A / 6A socket / switch). The UPS outlet is kept at deferent height and identification is provided to differentiate it normal power socket.

8.8.4 Design of Nodes: As per discussion with user of building, the number of computer with their location (nodes) are to be worked out and summarized. If WAP and USB device pair are to the installed at required location, to provide for operation of potable computer, laptop etc. Each WAP and USB device pair needs their separates UPS (500 VA) while nodel computers works on central UPS supply. A modular information outlet is provided at each node, which is connected to switch.

8.8.5 Design of Switch: The client computer (computer at node) is connected to server via switch, which has number of ports. As per international standard the switches are available in market with some fixed number of ports e.g. 8, 16, 24 ports are suitable.

These switches require a separate UPS of 500 VA. Provision of future extension of computer networking i.e. inclusion of more new computers say 20% margin is taken in to account in designing number of switches. The number of switches are so designed so that it feed nodes in same building, same floor in multistoried building so that so many individual connection wire are not supposed to cross the floor to avoid excessive cutting and aesthetic look point of view. A number of switches are found by dividing total number of nodes by port capacity of switch.

$$\text{Number of switches of N ports} = \frac{\text{Total number of nodes x 1.2}}{N}$$

A Jack Panel of same number of ports is provided with switch to facilitate connection from switch to node.

8.8.6 Design of Cable / Patch Code: A suitable optical fiber (OFC) for supplying data information from server to different switch in the building separate from server building, preferably to be laid in ground and UTP service cable to supply from server to switch inside same building and from switch to modular information outlet (MIO) to connect client computer. A six core OFC armored cable is selected for outside building while CAT (category) six service cable is preferred as UTP service cable. Line interface unit (LIU) boxes are required for connection of OFC and service cable. SC - SC cable patch code of same service cable is required for connection of switch to Jack panel. One three m and one 1m patch code is needed at each node to connect MIO with monitor and keyboard.

8.8.7 Design of Earthing: Suitable earthing is done to earth UPS or its control switch if it separately provided. Earth wire run along with submain from UPS to switch and UPS outlet (16 A power socket) at nodes.

8.8.8 Design of Internetwork: A suitable VSAT antenna with modular converter is provided for accessing internet by server, if their number of computer in LAN or a suitable modular connection is take from service provider such as BSNL, Relince etc.

8.9 Eastimating and Costing:

To prepare detailed estimate to find out the cost of installation of computer networking the following steps are taken.

8.9.1 Selection of Network: Proper network is selected after studying the building carefully, its scale of area (PAN,LAN,CAN), use of building, method of connection(wired or wireless), functional relationship(Peer to Peer, Client/server), network topology and requirement of internet work etc. After selection type of network, we mark number of nodes i.e. placese of computer to be installed and interconnected, on the plan of the building supplied by owner or construction agency. Location of WAP or USB device pair are also marked on plan as per their requirement. Then we design for different hardware or networking device required as follows.

8.9.2 Design of Server and its Room: Knowing the number and their location, type of computers to be installed, a proper server is selected with its complete recourses as per requirement of user of building after a length of discussion with user i.e. printer, scanner, VSAT with modular converter or broad band modem connection of service provider. Location of server room is also selected at central location nearer to video conference room, if exit as per 8.8.3 of this chapter.

8.9.3 Design of Hubs or Bridges: The number of hubs or brides are selected as per requirement if necessary.

8.9.4 Design of Network Repeater: If the distance of buildings in campus to be interconnected through server or length of interconnecting OFC cable is long, a network repeater is required and selected.

8.9.5 Design of Modular Information Outlet: Each node requires a MIO for connecting computer at node with server to receive data information. So the total numbers of MIO are taken as total number of nodes.

8.9.6 Design of Switches and Jack Pannel: A switch of sufficient number of ports is selected and total switches are designed as per 8.8.5 of this chapter the number of jack panels are taken same as number of switches.

8.9.7 Design of U.P.S.: In large network except in PAN, computer at nodes is supplied power from a common central UPS placed in server room. The capacity of UPS is designed depending on number of client computer (nodes) and their distances from server as per 8.8.3 of this chapter.

Small UPS (500VA) for each switch and WAP is required as they are not connected with central UPS. Total number of UPS outlet at each nodes are designed as equal to total number of client computers (node). A numbers of small UPS are equal to the total number of switches and WAP points.

8.9.8 Design of Data Cable and Patch Chord: Data cable from server to different switches is designed as per their locations and numbers whether inside or outside building to be burried in ground or provided in trenches on wall/ in floors or in conduit. Optical fiber cable of suitable size for outside building, cat 6 UTP cable for inside building are selected. Line interfaced unit are used for connection of OFC/UTP cable and patch chord are used for connection of jack panel switches.

8.9.9 Design of U.P.S. Wiring: Proper submain wiring with copper wire in concealed conduit is designed from server UPS to UPS outlet as per per total electrical load of computers and is done as per PWD specification. UPS submain wiring in video conference room nearer to the server with smaller size (1.5 sq mm) copper submain as compare to submain wiring(2.5 sq mm) for UPS outlet at distances.

8.9.10 Design of Earthing: Separate earthing is design for UPS as per 8.7.7 of this chapter and is done as per PWD specification.

8.9.11 Design of VSAT Modular Convertor: Service provider transmit the data information in shape of modulated carrier wave, via transmitter antenna with telephone signal, if the service provider also supply telecommunication signal for telephone set through telephone exchange. Receiver receives this data information through telephone broad band modem, to demodulate/encode data information or a VSAT antenna with modular converter from international companies on internet.

8.9.12 Detail of Measurment: After designed each device such as nodes, UPS outlet, switches, server, bridges, hubs, network repeater, MIO, UPS, LIU, wiring and data cabling, they are summarized in a table under detail of measurement to find there total quantities.

8.9.13 Analysis of Rates: Rates of each items to be executed are found after preparing proper analysis of rates on standard format after geting market rates or price list of manufacturer.

8.9.14 Bill of Quantity: A proper bill of quantity on standard format with complete specification of each items of computer networking is prepared to find the cost of installation for computer networking.

8.9.15 Abstract of Cost: Finally the abstract of cost is prepared including BOQ of the computer networking, the contingencies, and centage charges is prepared on standard format.

8.10 Video Conference:

In today's competitive world everybody want to make conversation, without being physically present with other, thus reducing travel cost the so demand of high qulity video conferencing(VC) is increasing.

Video conference service allows any number of participates to converse with each other regardless their location through video end point or personal computers. Video conference (VC) involves video and audio communication. So we can say VC is about connecting people. There is a provision that a VC subscriber can add two or more participates in a particular conference.

Any customer schedule their VC through the web video conferencing service, which can be availed by any customer through IP or ISDN interface. VC service are provided by many service provided e.g. BSNL. The customer of VC may have both dial-in and dial-out at participate on IP or ISDN. For VC one must have video and point with web camera or we can have PBX software of BSNL conference portal to avail VC facilities.

Many web based companies also provide free VC facility for which one has to be participants by opening web account on that site e.g. skipe or Google talk etc. The provision of TV, Stabilizer, web camera, microphone and amplifier etc. is made for arranging VC in addition to the correction to the obtained to the service provider. Accordingly estimation and costing can be done with intercom work if require to prepare the estimate and find out cost.

8.11 Intercom Telecommunication:

The device which is used to communicate with in building between persons sitting in any portion of building i.e. device used for internet communication in a building, inside is called intercom internal communication inside building, connecting to the outside world with telephone connection from exchange of any operator BSNL or Reliance etc.

So intercom is a essential need in big or multistoried building. If the building occupants are different owner e.g. residential building or commercial complex, with different entrepreneur e.g. shopping complex or office complex, a intercom facilities are provided by operator as BSNL or other. If the building pertains to one or more big offices or hotel etc., the owner provides telecommunication with intercom facility, taking single telephone connection. Every occupant in different room or portion with intercom facility may access other telecom through operator exchange, by extension facility of the owner as per desire. So arrangement for wiring for telecommunication is to be provided during construction of building. So it is also be a part of electrical installation to be delt with scope of the book. Several types of internal telecommunication (intercom) systems are available.

8.11.1 Basic Coponent of Intercom: Separate mini exchange or PABX is installed in the building and connection is provided to individual user through crone type distributor, if the building is owned by many occupant, then mini exchange or distributor are installed by telephone operator or service provider (BSNL etc.). If there is one or more big entrepreneur/office in building, then PABX is provided with connection to individual occupant by the entrepreneur with extension facility.

There are many type of PABX with different facilities.

(a) DID (direct inward dial) with PRI (Primary rate interface) facility provides direct number in telecom operator i.e. one can have number with level STD plus ABCDEF for direct dialing. So thousand numbers can be provided inside building. Each number can be dialed directly from outside building. As we have direct dial number we do not require any extension facilities.

(b) Many PABX has ADL card to extend the broad band connectivity to desired occupant (pair or connection) to use internet work with computer. SO telephone operator provides a modem to give separate PRI and ADSL link. So optical fiber cable is used for the purpose.

(c) Many PABX provides extension to extension call and is used as intercom or EPABX.

(d) PABX also have other facilities i.e. Music on hold, Call transfer secrecy, Automatic call transfer on other extension if main is busy or unattended, out going call restriction, Call pickup, Call line parking, Redial etc.

8.11.2 Cables : Different type of cables are required as per our requirement inside and outside building. Armored PIJF cable is required in outside building to the laid in ground, while PVC or unarmored PIJF cable is provided inside building. As we have to provide connection on floor wise in multi stored building and in different passage or galleries on same floor different pair cable is provided inside building (e.g. 50 pair, 20 pair,10 pair or 2 pair cable etc.). Unarmored cable run in concealed conduit in building. So parallel conduit run for different cable for voice (intercom) and data transfer (computer) in the building.

8.11.3 IDF or DP Box: To provide connection to individual occupant two pair cable is used. From EPABX individual two pair cable is not used instead fixed pair cable is brought from EPABX via to some node where DP is provided for two pair cable connection. These IFC and DP boxes are available with different fixed number of pair e.g. 200 pair IPF, 50 pair or 20 pair or 10 pair DP.

8.10.4 Home Intercom System: Wireless home intercom system provides audio/video communication link at a home 4-5 linked consol station in addition to 1-2 remote door well units. We use digital signal frequency in this system. There are two type home intercom available in market.

(i) Video intercom system:- The video intercom monitor about 2 ½" x 2 ½" are available inn color black with additional port to connect with 2-3 monitor (it is useful as door well).

(ii) Wireless intercom system:- The system is use to talk privately apart from as paging device to broadcast to all station (intercom) upto 1000 feet or so.

8.12 Estimating and Costing:

After selection of EPABX for requirement of user we can design for cable and DP boxes as per required number of intercom.

8.12.1 Detail of Measurment: After knowing the quantities we can summarized each item in detail of measurement.

8.12.2 Analysis of Rates: Proper analysis of rate are prepared after obtaining rates from market or manufacturer to find the rates of each item if their exit no schedule of rates.

8.12.3 Bill of Quantity: Bill of quantity is prepared for EPABX work on standard format describing with brief specification of each items, their quantities, rates and amount. Total of bill of quantity will give the cost of internal telecommunication work.

8.12.4 Abstract of Cost: Abstract of the cost is prepare on standard format including cost of intercom work as per BOQ, contingency, centage charges and other unforeseen items as lamp-sum basis.